Strategies for

MATHEMATICS INSTRUCTION AND INTERVENTION

6-8

Chris Weber

Darlene Crane

Tom Hierck

Solution Tree | Press

a division of
Solution Tree

555 North Morton Street
Bloomington, IN 47404
800.733.6786 (toll free) / 812.336.7700
FAX: 812.336.7790
email: info@solution-tree.com
solution-tree.com

Visit **go.solution-tree.com/mathematics** to access materials related to this book.

Printed in the United States of America

19 18 17 16 15   1 2 3 4 5

Library of Congress Cataloging-in-Publication Data

Weber, Chris (Chris A.)

   Strategies for mathematics instruction and intervention, 6-8 / by Chris Weber, Darlene Crane, and Tom Hierck.

      pages cm

   Includes bibliographical references and index.

   ISBN 978-1-936763-33-7 (perfect bound) 1. Mathematics--Study and teaching (Elementary) 2. Mathematics--Study and teaching (Middle school) I. Crane, Darlene. II. Hierck, Tom, 1960- III. Title. IV. Title: Strategies for mathematics instruction and intervention, grades 6-8.

   QA135.6.W434 2015

   510.71'2--dc23

                             2015011535

Solution Tree
Jeffrey C. Jones, CEO
Edmund M. Ackerman, President

Solution Tree Press
President: Douglas M. Rife
Associate Acquisitions Editor: Kari Gillesse
Editorial Director: Lesley Bolton
Managing Production Editor: Caroline Weiss
Production Editor: Tara Perkins
Proofreader: Elisabeth Abrams
Compositor: Rachel Smith
Cover and Text Designer: Rian Anderson

To the memory of Benjamin Bloom and his work on mastery learning—so simple and inspiring, and the predecessor to RTI.

———————————

To Michael Bartlett, an exceptional mathematics teacher whose life tragically ended too soon. In his review of an early draft of this book, he offered this comment for all mathematics educators to consider:

It's certainly revealing exactly how much work it will take to increase our success in math instruction . . . lots of systemic adjustments, overhaul, and expectations.

He approached mathematics instruction from the perspective of making it engaging for all students, and it's with this notion in mind that we offer this resource.

Acknowledgments

The support of Solution Tree—Jeff Jones, Douglas Rife, Shannon Ritz—is simply world class. Education and educators are infinitely better for the support of Solution Tree and its staff. The editorial team that brought this book to life—Tara Perkins, Caroline Weiss, Jessi Finn, Lesley Bolton, Rian Anderson, Sarah Payne-Mills, Rachel Smith, and Elisabeth Abrams—has our undying gratitude. The ideas represented within this book reflect our work with colleagues across North America. We gratefully and endlessly acknowledge their dedication to continuous improvement. A special thanks to the professional educators who have shared their wisdom, talents, and inspiration with us, especially those from Nanaimo-Ladysmith Public Schools and the State of West Virginia, and to Kirk Savage and Kevin Bird, authors of *The ANIE*, who provided feedback and examples for assessment.

Solution Tree Press would like to thank the following reviewers:

Sue Bergstrom
Math Interventionist
Adams-Friendship Middle School
Adams, Wisconsin

Melanie Doran
Algebra, Math Support,
   Interventionist
Gorham Middle School
Gorham, Maine

Megan Farrelly
Math Instructional Coach
Mark Twain Middle School
Alexandria, Virginia

Deanna Lomax
Math Teacher
Chief Joseph Middle School
Richland, Washington

Tamara Rasmussen
Oregon RTI
Roseburg, Oregon

Katie Scholl
Instructional Coach
Prairie Ridge Middle School
Ankeny, Iowa

Visit **go.solution-tree.com/mathematics** to access
materials related to this book.

Table of Contents

Chapter 3

Instructional Practices for Application-Based Mathematical Learning

Epilogue

The Promise and Possibility of Improved Mathematics Learning. 139

About the Authors

Chris Weber, EdD, is an expert in behavior, mathematics, and response to intervention (RTI) who consults and presents internationally. As a principal and assistant superintendent in California and Illinois, Chris and his colleagues developed RTI systems that have led to high levels of learning at schools across the United States.

In addition to writing and consulting, he continues to work with teachers and students within schools and school districts across Canada and the United States.

A graduate of the United States Air Force Academy, Chris flew C-141s during his military career. He holds a master's degree from California State University, San Marcos, and a doctorate of education from the University of California (Irvine and Los Angeles). He is a former high school, middle school, and elementary school teacher and administrator.

To learn more about Chris's work, visit http://chriswebereducation.com or follow him on Twitter @WeberEducation.

Darlene Crane is an author, speaker, and consultant with twenty-five years of experience in education. She is a vice principal for Nanaimo-Ladysmith Public Schools in British Columbia and an instructor at Vancouver Island University. Her professional experience also includes serving as an RTI specialist for the West Virginia Department of Education. Darlene has extensive general education and special education classroom teaching experience in addition to holding district leadership roles.

Darlene has trained teachers and administrators throughout North America on tiered instruction including examining the impact on culture, structure, and instruction within a school community. Her work has targeted both mathematics and reading in the elementary and middle grades. Darlene's expertise includes identifying targeted improvements for classroom instruction, designing effective intervention to accelerate student learning, and creating collaborative school cultures. Darlene holds a master's degree in special education from West Virginia University.

You can follow her on Twitter @DarleneCrane.

Tom Hierck has been an educator since 1983 in a career that has spanned all grade levels and many roles in public education. His experiences as a teacher, administrator, district leader, department of education project leader, and executive director have provided a unique context for his education philosophy that is built on a foundation of creating and sustaining a positive school culture.

Tom is an expert in assessment, behavior, curriculum design, and RTI. He understands that educators face unprecedented challenges and knows which strategies will best serve professional learning communities. Tom has presented to schools and districts across North America and internationally with a message of celebration for educators seeking to make a difference in the lives of students.

To learn more about Tom's work, visit http://tomhierck.com or follow him on Twitter @thierck.

To book Chris Weber, Darlene Crane, or Tom Hierck for professional development, contact pd@solution-tree.com.

The Rationale for RTI and Application-Based Mathematics

In today's world, economic access and full citizenship depend crucially on mathematics and science literacy.

—Robert Moses, Civil Rights Leader

Response to intervention (RTI) can be best understood as the teacher practices (instructional delivery, assessment, and student discipline) and typical processes (daily, weekly, or monthly routines, such as schedules and collaborative times) that have impact on and are impacted by the things we do in education. RTI is about using the knowledge, skills, and attributes of the members of a learning organization to positively impact the outcomes for all students. Schools and the educators found therein can make a difference in the lives of *all* students. RTI is not about waiting until we exhaust all other options and then hoping to deploy some sort of rescue mission or providing a last chance prior to the presumed academic failure; it is a proactive effort to ensure that students receive the supports they need as soon as they need them, whether the struggles occur with academic or behavioral expectations. The critical components of an effective RTI system are as follows.

- High-quality instruction and learning opportunities for all students
- Identification of students struggling to meet grade-level expectations
- Attention to the learning rates and performance levels of all students
- Increasing intensity of instruction and intervention based on identified student needs
- Data-informed decision making using the skills of the team to solve problems

In the domain of mathematics, this critical belief in the success of all students translates more acutely because schools have traditionally taught this content by algorithm or rule or by following a formula rather than by using an inquiry approach. This method has often resulted in "math memorization" proficiency that leads to students being less successful in everyday practical application of mathematics.

Implementation Lag of RTI for Mathematics

While schools have embraced the RTI model for reading and behavior, implementation of RTI for mathematics continues to lag (Buffum, Mattos, & Weber, 2009, 2010, 2012). Several factors may contribute to this lag in implementation. First, we have valued written and spoken language abilities over mathematics. The fact that the National Institute of Child Health and Human Development's *Report of the National Reading Panel* (2000) has impacted reading instruction to a greater extent than the National Mathematics Advisory Panel (NMAP, 2008) report has impacted mathematics instruction is one indication of this mindset. Many educators with whom we work across North America, including mathematics teachers, are unaware of the significant recommendations the NMAP report identifies with regard to mathematics instruction. Moreover, given that 80 percent of students identified as having a learning disability have deficits in reading (Lyon, n.d.; Shaywitz & Shaywitz, 2003), the Individuals With Disabilities Education Act's recommendation of RTI as a means for identifying students with a learning disability may also have influenced the earlier adoption of RTI for reading.

Another factor contributing to the lack of implementation of RTI for mathematics may be that teachers have not received sufficient professional support to address mathematics instruction and intervention practices. Many districts have invested more for professional development in the areas of reading and behavior than mathematics. Balanced, thorough, and practical resources and professional development for teaching mathematics have been limited compared to those that address reading and written expression. Assessment and intervention strategies for mathematics are also not as accessible to many teachers.

Although the implementation of RTI for mathematics has lagged behind reading, the need for RTI for mathematics is equally critical for today's students. While there are multiple, compelling reasons that RTI for mathematics has lagged in implementation, there are also multiple, compelling reasons for districts and schools to begin the journey with urgency. Simply stated, we must recognize the critical importance of mathematics and embrace the high-impact instructional approaches that are integral to the RTI framework. The current reality of our global context challenges these perceptions. Noted civil rights activists Robert Moses and Charles Cobb (2001) identify mathematics as a crucial element for economic access and full citizenship.

An International Perspective: The Critical Need for Improving Mathematics Achievement

Since the turn of the century, internationally respected organizations such as the National Council of Teachers of Mathematics (NCTM, 2000) have suggested reforms that contribute to a new understanding and appreciation of mathematical learning. Teachers and students throughout North America have begun to develop a greater appreciation of mathematics through learning experiences that include inquiry, collaboration, and hands-on learning. Mathematics lessons include a greater emphasis on exploring mathematics with manipulatives and the integration of collaborative learning experiences. However, as education researchers Barbara Nelson and Annette Sassi (2007) explain, these efforts alone will not sufficiently meet the needs of our students.

> Small-group work can miss the mark, working with manipulatives can fail to bring students into contact with the mathematical concepts these instructional tools are intended to illuminate, and enumeration of problem-solving strategies without discussion of their accuracy or effectiveness can deprive students of the opportunity for rigorous mathematical thinking. (p. 47)

For our students to develop the depth of mathematical knowledge we recognize as imperative, we must integrate rigorous, engaging instructional strategies into a balanced mathematics program—one that promotes conceptual understanding, competence with procedures, and effective, efficient problem solving. While there have been efforts to reform mathematics instruction, there is still a significant need for practical instructional strategies that allow our teachers to more fluidly respond to the diverse needs of today's learners. The education we provide our students must prepare them to be citizens in a global economy, to be both college and skilled-career ready. The perception that elements of written and spoken language define being a literate adult is no longer acceptable and does not serve our students well. While written and spoken languages are critical skills that our schools must remain committed to develop and enrich, mathematical reasoning and problem-solving skills are equally critical for today's students. NCTM and other international organizations that promote educational reform emphasize the critical importance of mathematical learning. The Partnership for 21st Century Skills, a national organization composed of businesses, educational organizations, and foundations that promote the advancement of college and career readiness for every student in the United States, recommends improving achievement in basic skill areas, including mathematics. These organizations have also identified interdisciplinary themes to weave into those areas to promote understanding at higher levels. Examination of these interdisciplinary themes (table I.1, page 4) clearly identifies the importance mathematics holds in the future both for our students and our culture.

Table I.1: Implications of 21st Century Interdisciplinary Themes for Mathematics Instruction

Interdisciplinary Theme	Description	Mathematics Connection
Global awareness	Using 21st century skills to understand and address global issues through working with individuals from diverse cultures and backgrounds	Problem solving, analysis of data
Financial, economic, business, and entrepreneurial literacy	Understanding the role of economics in society, making personal financial decisions, and using entrepreneurial skills to enhance workplace and career options	Making sense of part and whole, computation, algebraic concepts, and mathematical skills and understandings related to economics
Civic literacy	Preparing to participate in civic life by staying informed and understanding governmental processes and the local and global implications of civic decisions	Problem solving and analysis of data and financial elements
Health literacy	Understanding physical and mental health measures (including proper diet, nutrition, exercise, risk avoidance, and stress reduction) and using available information to make appropriate health-related decisions	Proportional reasoning, extending patterns, and probability in a variety of contexts
Environmental literacy	Demonstrating knowledge and understanding of society's impact on the natural world (including population growth, population development, and resource-consumption rate), investigating and analyzing environmental issues, and drawing accurate conclusions about effective solutions	Making sense of part and whole, computation, algebraic concepts, proportional reasoning, extending patterns, probability, and mathematical skills and understandings related to resources and the natural world

Educators worldwide realize that today's students face highly challenging problem-solving situations and global economic experiences. British educator John Abbott (2006), president of the 21st Century Learning Initiative, also expresses the importance of mathematical learning for our current and future reality: "The world, as it begins to dig itself out of its present ecological dark pit, will need vast numbers of mathematicians" (p. 1). These international perspectives reinforce our recognition that high levels of mathematical competency will unquestionably be a significant factor in our students' personal success and our global community's ability to solve some of the critical economic, environmental, and social challenges we face.

A Redefinition of the Mathematics We Teach and How We Teach It

In order to address these needs, educational systems across North America have invested significant time and resources in the revision of curriculum standards to promote a more rigorous understanding and application of mathematical concepts. Given international data related to students' mathematical performance, the need for increased rigor and cohesiveness for mathematics standards in North America is clear. Our extensive lists of curricular outcomes have impaired classroom teachers' abilities to ensure high levels of learning for students. The perceived need to cover mathematical *content* has compromised mathematical *learning*, which encompasses conceptual understanding, procedural fluency, and problem solving—skills that the final report of the NMAP (2008) identify as necessary.

Many U.S. states adopted the National Governors Association Center for Best Practices (NGA) and the Council of Chief State School Officers' (CCSSO) Common Core State Standards (CCSS, 2010a) as a first step in addressing these needs. Even those states that have not adopted the CCSS have reviewed and refined their standards to increase relevance, rigor, and alignment. In addition, many Canadian provinces, including British Columbia, Alberta, Manitoba, Nova Scotia, and Newfoundland, have revised their standards in order to promote a deeper understanding of mathematics (Alberta Education, 2014; Manitoba Education, 2013; Rapp, 2009).

Redefining the quantity and quality of what we teach through revisions to standards and outcomes is a critical first step to promoting understanding and application of mathematical learning. Teachers and students must have sufficient time to deeply and thoroughly explore and master critical mathematical concepts. However, the revision of standards alone will do little to increase actual student achievement. As educational systems identify more cohesive, rigorous standards relevant to today's learners, the investment of time and resources will need to shift to instructional approaches that connect standards to increased student achievement. The impact of next-generation standards such as the CCSS on students' mathematical achievement will be minimal unless district leadership teams, professional learning communities, and classroom teachers are able to reflect on and improve instructional practices for mathematics.

Adopting new standards without supporting the development of instructional practices that promote deeper understanding and fluid application of mathematics will have little impact on student achievement. For mathematics education, our reforms must include redefining both *what* and *how* we teach. We can't teach new things in old ways; we must strategically and intentionally integrate critical knowledge and skills within powerful, authentic learning experiences.

Instruction for 21st Century Mathematics

The CCSS and other 21st century curricular documents align and streamline mathematical standards as recommended by the NMAP report (2008), which represents the mathematical knowledge all students need to be college and career ready. The creators of the standards recognize that students learn at different rates and that some students need extra time and supplemental instruction to master the mathematics that prepares them to graduate from high school and pursue higher education or a meaningful career. Thomas R. Rudmik (2013), founder of the award-winning K–12 Master's Academy and College, states that we must prepare our students with skills to be future ready. For our teachers in grades 6–8, we must balance optimism with practicality. Implementation of RTI-based systems of support presents challenges different from those our elementary counterparts experience. While many of our elementary schools utilize RTI for mathematics and elementary teachers invest significant time and energy to improve student achievement, many of the students within our grades 6–8 classrooms have not yet benefited from rigorous RTI practices. As such, students may not possess a mastery of requisite mathematics skills. Middle school teachers describe classrooms that are composed of students struggling with basic number sense, early computation skills, place value, and part-whole relationships. Designing mathematical learning experiences that address these needs takes significant and systematic planning. Many of the elementary educators with whom we work admit to not feeling as confident teaching mathematics as reading; many of the middle-level teachers we work with admit to not being as comfortable addressing these needs in basic mathematics as part of their grade-level content.

The diversity in student skill acquisition is not a new challenge for our schools. However, historically, students needing more support to master mathematics received instruction at the expense of the learning that was occurring in mainstream classes. Other students experienced watered-down mathematics courses that relied too heavily on computational, procedural practices as opposed to courses involving the integration of problem-solving experiences (Darling-Hammond, 2010). These traditional approaches to supporting students struggling with mathematics have done little to promote accelerated learning or close the achievement gap and are explicitly discouraged within the CCSS narrative. In contrast, the practices and processes of RTI involve systematic responses for students who need more time and support to master prioritized mathematical outcomes that promote accelerated learning to close achievement gaps (Buffum et al., 2009, 2010, 2012; Kanold & Larson, 2012; NGA & CCSSO, 2010b).

Implementing an RTI framework for mathematics allows schools to respond to student needs more systematically and proactively. Within an RTI framework, students engage in the rich, deep learning experiences of Tier 1 instruction while receiving the necessary, targeted interventions to close achievement gaps and to accelerate learning. We can predict that Tier 1 will not be sufficient for some students to demonstrate mastery and that periodic, need-based supports requiring a little more time and using alternative approaches will be needed. This additional time and these alternative approaches define Tier 2. Intensive, individualized interventions are provided for students with significant mathematical deficits (those who are multiple years behind in foundational prerequisite skills). The goal of Tier 3 intervention is to accelerate learning through diagnostically driven support to close achievement gaps, not to provide separate but unequal remedial mathematics instruction. Thus, RTI shifts a school's response to struggling mathematics learners from a remediation model that too often masks learning needs to targeted interventions that promote accelerated learning and higher levels of achievement (Buffum et al., 2009, 2010, 2012). A school's RTI process must allow for Tier 1, Tier 2, *and* Tier 3.

Response to intervention continues to be identified as one of the most effective instructional frameworks to ensure student achievement (Hattie, 2012), empowering educators to continuously improve instruction and assessment to support students' mastery-level learning. The collaboration inherent in professional learning communities (DuFour, DuFour, Eaker, & Many, 2010) is fundamental and foundational to RTI's success in systematically responding to student needs, directly and positively impacting student achievement. When educators work together to identify essential learning outcomes, monitor students' mastery of the *must-knows*, and provide timely, targeted interventions for students who demonstrate a need, schools can promote the critical mathematical knowledge students will need to apply throughout their lives.

We recognize the need for our students to be more confident with mathematical competencies and to be better prepared for high school mathematics courses; we also recognize the need to increase our understanding of mathematics and to increase our repertoire of evidence-based instructional and assessment practices. Teacher mathematical knowledge and evidence-based instructional practices have significant impacts on student learning (Slavin & Lake, 2007). A teacher's understanding of mathematics and how various mathematics concepts interconnect enhances his or her ability to respond fluidly to student learning. The understanding of mathematics influences a teacher's ability to develop effective questions to clarify thinking, challenge misconceptions, and provide real-time formative feedback. Increasing teachers' understanding of mathematics allows them to communicate more effectively and coherently with learners (Faulkner & Cain, 2013). Yet, while significant investments have been made to ensure elementary teachers are equipped with an understanding of the elements of reading, such as phonological awareness, phonics, fluency, vocabulary, and comprehension, equal investments have not been made in the areas related to early numeracy, despite clear evidence that increasing teacher knowledge of mathematics has positive impacts on student performance (Hill, Rowan, & Ball, 2005). While teachers' understanding of early numeracy concepts and procedures has not historically been as robust at the middle levels (grades 6–8), the needs of vulnerable students necessitate that middle-level teachers develop these understandings and incorporate related instructional practices into middle school learning experiences.

While developing conceptual understanding of mathematics and the interconnected nature of the discipline is vitally important, we also accept that richer learning experiences are necessary to engage today's students in new and innovative ways. Instructional practices that engage students in constructing meaning, drawing connections, developing procedural fluency, and applying mathematics are critical to increasing student achievement. We must also explore the most effective integration of the rich digital learning environments that are available for our students. We must provide teachers with practical instructional strategies and frameworks that scaffold student learning. As an educational community, we have recognized that greater focus and coherence of curriculum and standards are imperative. Now, we must recognize and address the need for increased teacher knowledge and attend to *how* we teach mathematics. We propose that with this greater focus, we shift our mathematics learning to what we identify as an *application-based* learning model through which students will master mathematics as the questions "Why do I need to know this?" and "How will I use this?" are answered. Application-based models bring conceptual and procedural understandings to life; they systematize problem solving and critical thinking within the classroom and the learning experience. In addition, application-based mathematics promotes learning that can be applied across content areas and in diverse problem-solving experiences.

How do we support teachers on this journey? We know our teachers are dedicated, that they search for the best resources, and that they exhaust all options to meet student needs. We also know that they require support as they revise and improve their practices. Teacher education programs alone will not sufficiently meet the needs of our students in a timely manner, though teacher-education programs do play a crucial role. School systems must support teachers in developing a deeper understanding of the mathematics they teach and provide them with practical instructional and intervention practices to meet the diverse needs of their learners. We must also provide our teachers with practical, implementable, researched-based resources.

This book strives to serve as an interconnected resource for teachers, strengthening their understanding of critical mathematics and equipping them with instruction, assessment, and intervention strategies to meet the complex, diverse needs of students. We specifically address five broad areas that can serve as guideposts for individual teachers, collaborative teams, PLCs, and teacher preparatory programs. Our goals are to provide:

1. **An overview of the most essential mathematical standards that should represent the focus of instruction and intervention**—Chapter 1 introduces a practical process for identifying essential learning outcomes that teacher teams can use to facilitate collaboration when identifying essential learning outcomes. The chapter also includes sample lists of essential outcomes derived from the most current recommendations.

2. **Sample units of instruction that are scoped and sequenced to viably lead all students to deep understanding of essentials (need-to-know information) and the authentic assessments that can be collaboratively designed, commonly administered, and collectively analyzed to comprehensively respond to student needs**—Chapter 2 provides a practical description of how to develop units of instruction that promote the interrelated nature of mathematics.

3. **Samples and explanations of evidence-based instructional practices that promote conceptual understanding, procedural understanding, and application for high-quality Tier 1 mathematics instruction**—Chapter 3 provides descriptions and examples of mathematical practices and instructional strategies that promote *application-based mathematics learning*—learning mathematics in a way that fosters the application of mathematics across content areas and in diverse problem-solving opportunities.

4. **A description of the prerequisite foundational and conceptual understandings and mathematical processes required of students with significant deficits in foundational skills, in need of Tier 3 interventions**—Chapter 4 provides an overview of these foundational skills and suggested intervention strategies, identifying the foundational skills

that students are missing and providing effective, targeted intervention critical for addressing achievement gaps.

5. **Recommendations for assessments that inform teaching and learning**—Universal screening assessments, diagnostic interviews, and progress-monitoring tools inform intervention strategies that promote student learning of the identified essential mathematical concepts. Chapter 5 provides explanations and examples of assessments needed for the RTI process and intervention strategies for the identified essential learning outcomes.

The epilogue outlines a vision for implementing RTI for mathematics.

We have organized this book to align with the work of teacher collaborative teams within a PLC and RTI model. Because of our ongoing work in schools and with teachers, we know that addressing these goals is necessary to support the implementation of response to intervention in grades 6–8. Our intention is to develop a resource that will provide the research (the why), the strategies (the how), and the resources (the what) that enable teachers to approach mathematics instruction and intervention as confidently as they do reading and written expression.

Chapter 1

Prioritized Content in Mathematics

There is, perhaps, no greater obstacle to all students learning at the levels of depth and complexity necessary to graduate from high school ready for college or a skilled career than the overwhelmingly and inappropriately large number of standards that students are expected to master—so numerous, in fact, that teachers cannot even adequately *cover* them, let alone effectively teach them to mastery. Moreover, students are too often diagnosed with a learning disability because we have proceeded through the curriculum (or pacing guide or textbook) too quickly; we do not build in time for the remediation and reteaching that we know some students require. We do not focus our efforts on the most highly prioritized standards and ensure that students learn deeply, enduringly, and meaningfully (Lyon et al., 2011). In short, we move too quickly trying to cover too much.

Prioritized Standards

We distinguish between prioritized standards and supporting standards. We must focus our content and curriculum, collaboratively determining which standards are *must-knows* (prioritized) and which standards are *nice-to-knows* (supporting). This does not mean that we won't teach all standards; rather, it guarantees that all students will learn the prioritized, must-know standards. To those who suggest that *all* standards are important or that nonteachers can and should prioritize standards, we respectfully ask, "Have teachers not been prioritizing their favorite standards in isolation for decades? Has prioritization of content not clumsily occurred as school years conclude without reaching the ends of textbooks?" Other colleagues contend that curricular frameworks and district curriculum maps should suffice. But we ask, "Will teachers feel a sense of ownership if they do not participate in this process? Will they understand *why* standards were prioritized? Will they stay faithful to first ensuring that all students master the *must-knows*, or will teachers continue, as they have for decades, to determine their own priorities and preferences regarding what is taught in the privacy of their classrooms?"

Focusing content and curriculum also requires that we collaboratively create clarity—that all teachers have the same interpretation of the meaning of standards. The teacher teams that will provide the instruction that ensures that students master the standards must interpret the *educationese* in which standards are typically written. Processes that guide instruction and instructional decisions while also informing the selection of common formative assessment items can support teams in this practice. Education author and presenter Mike Mattos and his teacher teams at Pioneer Middle School in Tustin, California, have pioneered such a process, encapsulated in the Essential Standards Chart (Buffum et al., 2009, 2012). Once standards are prioritized and clarified, teacher teams must determine the number of must-know standards that can be viably taught so that all students can master them. This step often involves teams flexibly placing standards within maps or calendars that ultimately define units of study. Optimally, the process of mapping academic content is also conceptually organized and vertically articulated, from grade to grade or course to course. The word *fidelity* continues to challenge our decisions when concentrating instruction. While we recognize the benefits of, and necessity for, curricular materials, we believe that fidelity to standards and student needs is the very best way of ensuring a guaranteed, viable curriculum (Marzano, 2003). Fidelity to the identified prioritized outcomes and student needs is much more impactful on student learning than fidelity to curricular materials and resources. We are overwhelming educators and students with the standards that fill most sets of state curriculum and textbooks. While this stress directly affects the breadth of student learning, it also impacts the depth of mastery at which learning can occur. An extensive quantity of research advocates depth over breadth. See, for example, Trends in International Mathematics and Science Study (TIMSS) reports (http://nces.ed.gov/timss) and the works of William Schmidt (2004) and Robert Marzano (2003). Until we address the lack of focus on the outcomes that we expect for all students, the high levels of learning that we expect in order to ensure that students graduate ready for college or a skilled career will elude us. In addition, as we give consideration to our students' continual progression, we must identify those prioritized standards that promote successful transitions across grade levels.

Concepts, Procedures, and Applications

The need for focus and clarity is greater in mathematics than it is in any other content area. The implications of favoring breadth over depth in mathematics are exacerbated given that the content we have attempted to cover has been focused on procedures. While procedural competency and fluency are critical, they must be balanced with opportunities to build conceptual understanding and apply mathematics to real-world situations. Concepts, procedures, and applications are mutually reinforcing (Rittle-Johnson, Siegler, & Alibali, 2001). Mathematics knowledge is interrelated and cross-curricular. A perhaps unintended consequence of the standards movement, dating back to *A Nation at Risk* (National Commission on Excellence in Education, 1983), has been to seemingly communicate to teachers and students that learning outcomes are disconnected. By numbering and categorizing standards, we may be sending the message that the knowledge represented within a standard is unique. By categorizing standards within domains, we may have missed opportunities to connect ideas. For students to gain deep mastery of mathematics—deep enough for them to retain understanding from year to year, deep enough for them to apply their knowledge in unique situations, and deep enough for them to justify their solutions and critique the reasoning of others—we must design instructional units that reinforce this connectedness. For example:

- The numeracy concept of conservation is connected and foundational to the decomposition of numbers.

- The decomposition of numbers is connected and foundational to the base-ten number system.

- The base-ten number system is connected and foundational to multidigit addition and subtraction.

- Multidigit addition, subtraction, multiplication, and division are connected and foundational to linear measurement, area, perimeter, surface area, and volume.

- Linear measurement is connected and foundational to the measurement of 2-D objects.

- Fractions are ratios, ratios can represent rates, and constant rates of change are synonymous with the slopes of lines.

- Proportions relate to scale drawings, which relate to similar triangles, which relate to the slope of a line.

What Is Conceptual Understanding?

Conceptual understanding addresses critical questions, Why does this mathematics work, and why is this mathematics true? Mathematics educator, author, and consultant John Van de Walle (2004) writes that "conceptual knowledge of mathematics consists of logical relationships constructed internally and existing in the mind as a network of ideas. . . . By its very nature, conceptual knowledge is knowledge that is understood" (p. 27). Conceptual understanding can help students make meaning of the algorithms that they are often asked to manipulate in their development of procedural understanding and connect these algorithms to previously assumed disparate content domains. As we describe later in this chapter, while conceptual understanding need not absolutely precede procedural understanding, an emerging conceptual understanding leads to the ability to retain and apply skills and concepts. Authors Bethany Rittle-Johnson, Robert Siegler, and Martha Wagner Alibali (2001) define conceptual knowledge "as implicit or explicit understanding of the principles that govern a domain and of the interrelations between units of knowledge in a domain" (pp. 346–347). A sound conceptual understanding leads teachers and students to answer the *how* and the *why* questions behind mathematical topics. Why do we regroup during multidigit addition and subtraction? How are various quadrilaterals related? A conceptual understanding answers these questions; it includes knowledge of relationships and internal connections within mathematics. When students are given the opportunity to grapple with topics and when teachers guide students in constructing knowledge, conceptual understanding—which can deepen procedural understanding—emerges.

The NCTM's (2006) *Curriculum Focal Points for Prekindergarten Through Grade 8 Mathematics* defines conceptual focal points as a

> connected, coherent, ever expanding body of mathematical knowledge and ways of thinking. Such a comprehensive mathematics experience can prepare students for whatever career or professional path they may choose as well as equip them to solve many problems that they will face in the future. (p. 1)

The final report of the NMAP (2008) states that "understanding core concepts is a necessary component of proficiency with arithmetic and is needed to transfer previously learned procedures to solve novel problems" (p. 26). Mathematics curriculum and instruction in too many classrooms have been too heavily focused on procedures. While achieving a greater balance is absolutely necessary, we must not allow the pendulum to swing abruptly in the other direction; we must develop students' procedural *and* conceptual competencies as well as ensure that students are provided rich opportunities to apply the ideas for which they are developing these understandings.

What Is Procedural Competency?

Procedural competency is a critically important companion to conceptual understanding. Rittle-Johnson and colleagues (2001) define procedural knowledge "as the ability to execute action sequences to solve problems" (p. 346). Van de Walle (2004) defines procedural knowledge as "the rules . . . that one uses in carrying out routine mathematical tasks and also the symbolism that is used to represent mathematics" (p. 27).

We have blended Rittle-Johnson's and Van de Walle's definitions as follows: Since the beginning of the discovery of formal mathematics, thinkers have developed more efficient ways of solving problems. Whether formulas, rules, algorithms, or techniques, the best procedures are mathematically sound and universally applicable shortcuts. Procedures are often understood as describing *what* to do and can perhaps best be categorized as efficiencies. All procedures, however, rest on mathematical concepts.

A powerful and significant element of procedural competency is that students recognize that more than one procedure or algorithm can be used to solve the same problem. Procedural competence must include the examination of multiple solutions to problems and an evaluation of the effectiveness of various strategies.

We must collaboratively and intentionally plan for the teaching and learning of both procedural and conceptual understandings within our instructional units. The NMAP (2008) states,

> Debates regarding the relative importance of conceptual knowledge, procedural skills (e.g., the standard algorithms), and the commitment of addition, subtraction, multiplication, and division facts to long-term memory are misguided. These capabilities are mutually supportive, each facilitating learning of the others. Conceptual understanding of mathematical operations, fluent execution of procedures, and fast access to number combinations together support effective and efficient problem solving. (p. 26)

Rittle-Johnson and colleagues (2001) conclude that "conceptual and procedural knowledge develop iteratively, with increases in one type of knowledge leading to increases in the other type of knowledge, which triggers new increases in the first" (p. 346). Recognizing the critical importance of this relationship allows teachers as well as students to flexibly identify, understand, and value varied approaches in problem-solving situations. Ensuring that students can explain and justify the use of algorithms and procedures must be part of the development of procedural competency. Explicitly modeling why these algorithms are mathematically sound and ensuring that students can articulate why they are sound are critical. Explicitly modeling *how* the steps of an algorithm are employed in an organized manner is equally critical. Written and visual guidance, at times through the use of mnemonics, may be required as a temporary scaffold while students develop proficiency. A powerful and significant element of procedural competency is that students recognize that they can use more than one procedure, strategy, or algorithm to solve the same problem. Procedural competency must involve students examining multiple solutions to problems and evaluating the effectiveness of various strategies.

Finding the time within the existing number of school days for both conceptual and procedural teaching and learning will require us to be focused on the most highly prioritized mathematics content within a grade level—what the CCSS call critical areas of focus. Finding that time will also require that we plan instruction that makes explicit connections between mathematical ideas.

Concepts Versus Applications

Many educators use the terms *concepts* and *applications* synonymously; however, we see them as distinct components of a balanced, comprehensive understanding of mathematics. Simply stated, a conceptual understanding addresses why the mathematics works; we can apply conceptual understandings, as well as procedural knowledge, to real-world, problem-solving situations.

The NMAP (2008) states that "to prepare students for Algebra, the curriculum must simultaneously develop conceptual understanding, computational fluency, and problem-solving skills" (p. xix). Students must be able to *apply* mathematics and model and solve real-world problems through mathematical thinking. The most authentic and important applications or problem-solving tasks will be those that can be solved using multiple approaches and that may have multiple solutions. To help classrooms reach this level of depth, we will likely need to begin by teaching students to persevere, think critically, and work collaboratively.

We might visualize the interconnectedness of the three realms of conceptual understanding, procedural competency, and application of mathematics to real-word problems with the schematic featured in figure 1.1.

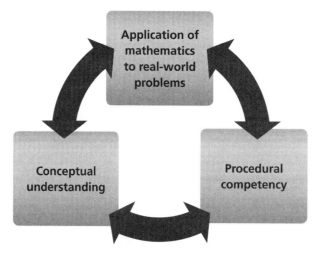

Figure 1.1: Conceptual understanding and procedural competency applied to real-world problems.

In our workshops, teachers often seek support regarding how concepts, procedures, and applications can be blended. Let's take a look at how concepts, procedures, and applications complement one another during the teaching and learning of mathematics in the example featured in figure 1.2. This example examines the addition and subtraction of integers. Teachers and students may first represent integers concretely with integer chips, building on this understanding to combine or add the chips, thus modeling the problem, 4 + (–6). Students create zero pairs to find a solution to 4 + (–6).

Figure 1.2: Visual representation of operations with integers.

Next, teachers and students may represent and display the sum by using a number line (see figure 1.3) to first identify the positions of integers before adding.

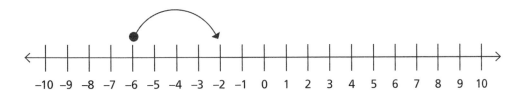

Figure 1.3: Using a number line to add and subtract integers.

Then, teachers and students solve the problem procedurally. The class may start by reinforcing the principle that subtraction is simply adding the inverse, perhaps by using a mnemonic like KiSS (Keep it, Switch, Switch).

For example, in a variation on the preceding problem, 4 + (–6), students *keep* the sign of the first number (4), *switch* the sign of the operation (from + to –), and *switch* the sign of the second number (from –6 to 6). The problem would then read, 4 – 6.

Another mnemonic, this one a song, could be used to add any integers, positive or negative.

Same sign, add
Different sign, subtract
Keep the sign of the higher number (the number with the greater absolute value)
Then you'll be exact

In the problem 4 + (–6), the numbers (4 and –6) have different signs, so, as the song indicates, the number would be subtracted. The higher number—the number with the greater absolute value—is –6, so the sum is negative. Teachers would wisely compare the solutions gained through the use of integer chips, number lines, and the mnemonic-driven procedures. As mentioned earlier, this use of mnemonics may be temporarily required as a scaffold while students develop proficiency.

As conceptual understandings and procedural competencies are developing, or as a unit-opening activity, teachers and students may investigate contexts in which they could apply the addition of integers. For example, we could present students with the following problem.

The temperature in Nome, Alaska, this morning was –5 degrees Fahrenheit. By the afternoon, the temperature had increased 14 degrees. What is the temperature in the afternoon?

Students could ... es, including the three described p ... a thermometer, essentially a verti...

Amount Of content we teach is greater

Depth Ver...

The 1990s intr ... rase "mile wide, inch deep." Edu ... 2004) coined the phrase during th ... n of the TIMSS. After controlling ... e performance of U.S. students, sp ... th-grade students compared to fou ... nificant factor in the poor perform ... ics topics covered in classrooms. H ... empt to master an average of seventy-eight topics per K–8 grade level, ... dents in Germany and Japan (both countries in which students significantly outperform U.S. students, after controlling for variables) attempt to master an average of twenty-three and seventeen topics, respectively, per K–8 grade level. The average length of a fourth-grade U.S. mathematics textbook is 530 pages; the average length of international textbooks is 170 pages (Gonzales et al., 2008). Depth of understanding has been sacrificed for breadth of coverage for too long, and deep student learning and retention of knowledge have suffered. By the way, for those who still maintain that we must teach everything because everything will be assessed, the worst way to prepare students for a test that assesses everything is to try to cover everything. Instead, let's ensure students master procedures, concepts, and applications for a viable quantity of content—content that we collaboratively deem most critical. Focusing on the priorities will serve all students in multiple ways—test scores will be higher than ever, students will be more prepared than ever for subsequent grade levels, and they will know how to think and problem solve, skills that will serve them for the rest of their lives. A focus on priorities is not simply better for students at risk. Even students currently passing the current high-stakes tests are ill prepared for college or a skilled career (Northwest Evaluation Association, 2011).

Author and professor James Hiebert (1999) studies the same TIMSS evidence from a different point of view. Watching videos of classrooms in action that were a part of the study, Hiebert notes that the nature of classroom practices in more successful countries is significantly different than in the United States. Students spend more time deeply investigating topics, possibly because there are fewer topics to address. If we're going to teach in the ways that Hiebert insists are necessary for true mathematics learning, learning that will allow students to apply their learning in meaningful contexts, then we, too, will need to devote more time to key topics.

This theme is reinforced in *Adding It Up* (Kilpatrick, Swafford, & Findell, 2001), which notes five attributes associated with deeply proficient mathematics learning.

1. Conceptual understanding (comprehension of mathematical concepts, operations, and relations)

2. Procedural fluency (skills in carrying out procedures flexibly, fluently, and appropriately)

3. Strategic competence (ability to formulate, represent, and solve mathematical problems)

4. Adaptive reasoning (capacity for logical thought, reflection, explanation, and justification)

5. Productive disposition (habitual inclination to see mathematics as sensible, useful, and worthwhile, coupled with a belief in diligence and one's own efficacy). (p. 116)

We doubt that a single teacher at any grade level can honestly state that achieving these goals is possible when attempting to cover all standards or complete all lessons in a text. The NMAP (2008) recommends focus and coherence in mathematics curriculum that avoid continually revisiting topics year after year without closure and allow for the engagement of topics to an adequate depth. In fact, the terms that authors of next-generation standards such as the CCSS mention most often are *focus* and *coherence*. Students' success at deeply mastering mathematics will require that teacher teams focus units of instruction on the most highly prioritized content. Furthermore, the scope and sequence within and between grade levels require coherence between related mathematics topics. Focused, coherent units of instruction provide the foundation of all of RTI.

RTI starts with Tier 1, and Tier 1 starts with carefully and completely defining key core content (Buffum et al., 2009, 2010, 2012). Why must we define key core content?

- Learning will only be guaranteed and viable if teams of educators define it clearly (Marzano, 2001).

- There are too many standards, even in light of the NGA and CCSSO (2010a) Common Core State Standards (DuFour & Marzano, 2011; Schmoker, 2011).

- Educators and students must know what mastery looks like, so that instruction can match these expectations and teams of educators can plan backward (McTighe & Wiggins, 2012). Typically, mastery is represented by the end-of-unit assessments that teacher teams collaboratively craft. Once crafted, instruction can be planned backward from the end to the beginning of the unit so that students are well prepared to demonstrate mastery at the unit's conclusion.

- The better our understanding of content and the more precisely we unwrap standards, the better our assessments of student learning, which leads to more accurate identifications of students in need of extra support, more diagnostic analyses of specific areas of need, and more targeted interventions (Buffum et al., 2009, 2010, 2012).

- We cannot intervene and provide more time and differentiated supports on all standards for students at risk; identifying the most highly prioritized standards helps determine the focus of interventions (Buffum et al., 2009, 2010, 2012).

In the remainder of this chapter, we describe how teams might prioritize standards. We also provide samples of the possible scopes and sequences of standards that allow for depth of learning within sixth- through eighth-grade classrooms.

How to Prioritize Standards

Even next-generation standards (for example, the CCSS, Texas Essential Knowledge and Skills, Virginia's Standards of Learning, and British Columbia's

Prescribed Learning Outcomes) represent a broad set of learning targets for students to master at an appropriate level of depth and complexity—a fact the authors of the standards recognize and communicate. A learning target is any achievement expectation for students on the path toward mastery of a standard. Targets clearly state what students should know and be able to do. Within the CCSS, for example, an introduction that describes the topics that should represent critical areas of focus where instructional time should be prioritized precedes each grade level's standards. An agreed-on set of prioritized standards is critical to ensuring a guaranteed, viable curriculum, but the biggest impediment to implementing standards is the *number* of standards (Marzano, 2001). Prioritizing standards is therefore a must for teams. A number of steps can assist teams in collaboratively identifying the standards and learning targets that should be prioritized. These steps are an extension of the work of Reeves (2004) and Ainsworth (2003a).

1. List the standard in a standards column.

2. If necessary and desired, list more specific learning targets.

3. Determine whether the standard is a prerequisite for the next grade level.

4. Determine whether the standard is important within other content areas.

5. Determine whether the standard is a critical life skill for college and skilled-career readiness.

6. Determine whether the standard is heavily weighted on high-stakes tests (if applicable).

7. Determine whether the standard is an area of need for your students.

Then use the following six levels of Bloom's Revised Taxonomy (Anderson & Krathwohl, 2001) to identify the level of this standard.

1. **Remembering:** Recalling information (recognizing, listing, describing, retrieving, naming, finding)

2. **Understanding:** Explaining ideas or concepts (interpreting, summarizing, paraphrasing, classifying, explaining)

3. **Applying:** Using information in another familiar situation (implementing, carrying out, using, executing)

4. **Analyzing:** Breaking information into parts to explore understandings and relationships (comparing, organizing, deconstructing, interrogating, finding)

5. **Evaluating:** Justifying a decision or course of action (checking, hypothesizing, critiquing, experimenting, judging)

6. **Creating:** Generating new ideas, products, or ways of viewing things (designing, constructing, planning, producing, inventing)

Finally, use the following four levels of Webb's (1997) Depth of Knowledge to identify the level of this standard.

1. **Recall:** Requires recall or recognition of a fact, information, concept, or procedure

2. **Basic application of skill or concept:** Involves use of information and conceptual knowledge, selecting appropriate procedures, following two or more steps with decision points along the way, solving routine problems, and organizing and displaying data

3. **Strategic thinking:** Requires reasoning and developing a plan or sequence of steps to approach a problem and some decision making and justification; is abstract and complex, often involving more than one possible answer

4. **Extended thinking:** Requires an investigation or application to the real world; requires time to research, think, and process multiple conditions of the problem or task and nonroutine manipulations across disciplines and content areas and from multiple sources

Examples of Prioritized Standards, Grades 6–8

In this section, we recommend prioritized standards in each grade span. We will first summarize next-generation standards from the beginning of the 21st century from various states and provinces. Next, we will share our recommendations based on the actual work of teacher teams with whom we have worked. These teams have committed to making sense of and designing focused sets of standards that represent a guaranteed, viable curriculum. Furthermore, these recommendations strive to include coherence both within (horizontal) and between (vertical) grade levels. Our recommendations are meant only to generate dialogue. We strongly encourage each local school system to independently analyze, interpret, and prioritize next-generation standards. Such decisions are then contextualized to local realities, and the process of collaboratively prioritizing standards will lead to a significantly deeper level of understanding and a greater sense of ownership. First, we share the prioritized standards determined by the NGA and CCSSO (2010c). They use language that suggests that instructional time should focus on critical areas. While there are approximately three to four dozen standards in each grade level, NGA and CCSSO have prioritized a more focused set of topics or critical areas. Table 1.1 lists these for grades 6–8.

Table 1.1: CCSS Prioritized 6–8 Topics

Grade 6	Grade 7	Grade 8
1. Interpret ratios and solve rate problems.	1. Solve proportion and percent problems.	1. Recognize, generate, interpret, and analyze linear equations and their graphs.
2. Multiply and divide with fractions.	2. Find and interpret unit rates, including those that describe linear relationships in the first quadrant of the coordinate plane.	2. Solve and interpret equations and systems of linear equations.
3. Compute with rational numbers (except integers).		3. Understand and describe functions.
4. Make sense of integers on a number line and in the coordinate plane.	3. Represent quantities in various forms—decimals, fractions, and percents.	4. Interpret and generate transformations.
5. Evaluate and interpret expressions.	4. Compute with all rational numbers.	5. Solve problems involving transversals and parallel lines.
6. Solve and interpret simple equations and inequalities.	5. Solve and interpret multistep equations and inequalities.	6. Interpret and apply the Pythagorean theorem.
7. Find and interpret measure of center or variation.	6. Compute the area and circumference of circles.	
8. Compute surface areas using nets.	7. Compute surface areas and volumes of prisms.	
9. Compute volumes of rectangular prisms.	8. Generate and draw inferences from data sets.	

The State of New York (EngageNY, 2014), among other states, has proposed prioritized scopes and sequences for next-generation standards. These guides can inform the work of teacher teams in prioritizing standards. Table 1.2 lists these for grades 6–8.

Table 1.2: State of New York Prioritized 6–8 Scopes and Sequences

Grade 6	Grade 7	Grade 8
1. Ratios and unit rates	1. Proportional relationships and percents	1. Scientific notation
2. Operations with rational numbers (except integers)	2. Operations with rational numbers	2. Congruency and similarity
3. Expressions and equations	3. Expressions and equations	3. Linear equations
4. Surface area and volume	4. Statistics	4. Functions
5. Statistics	5. Geometry	5. Pythagorean theorem and irrational numbers

In grades 3 and above, next-generation assessments will measure student learning of the next-generation standards. The organizations that are creating these assessments have prepared frameworks that define the relative weight of various standards at each grade level. Table 1.3 details how one of these organizations, the Partnership for Assessment of Readiness for College and Careers (PARCC), prioritizes clusters of standards for grades 6–8.

Table 1.3: PARCC's Prioritized Clusters of 6–8 Standards

Grade 6	Grade 7	Grade 8
1. Interpret ratios and solve rate problems.	1. Interpret and solve proportions.	1. Evaluate radicals and expressions with integer exponents.
2. Multiply and divide fractions.	2. Compute with rational numbers.	2. Interpret, produce, and solve linear equations and systems of linear equations.
3. Compute with rational numbers (except integers).	3. Manipulate expressions.	3. Interpret, compare, and model with functions.
4. Evaluate expressions.	4. Evaluate expressions and solve equations.	4. Interpret and produce congruent and similar shapes.
5. Interpret and solve one-variable equations and inequalities.		5. Interpret and apply the Pythagorean theorem.
6. Interpret and write equations that directly relate two variables.		

Now that we have provided examples of prioritized 6–8 standards from the CCSS and policy groups, we share our own prioritized learning targets in table 1.4. We have several objectives in referencing the prioritized lists of topics, domains, and clusters of other organizations when crafting our own lists.

- We want to communicate priorities in a language that teachers will understand, leading to a smoother transition from standard to classroom learning target.

- We want to incorporate our knowledge of student needs and learning progressions while recognizing that our contexts may not match the realities of others.

- We want to complete the process ourselves to gain a deeper understanding of the standards and to feel a sense of ownership over our decisions.

Table 1.4: Weber, Crane, and Hierck Prioritized 6–8 Standards

Grade 6	Grade 7	Grade 8
1. Interpret and solve rate problems. 2. Compute with rational numbers (except integers). 3. Make sense of integers and other rational numbers on a number line and integers in the coordinate plane. 4. Evaluate, solve, and interpret expressions and simple equations and inequalities. 5. Compute surface areas and volumes of rectangular prisms.	1. Solve proportion and percent problems. 2. Represent quantities in various forms—decimals, fractions, and percents. 3. Compute with all rational numbers. 4. Solve and interpret multistep equations and inequalities. 5. Compute the area and circumference of circles and the surface areas and volumes of prisms.	1. Recognize, generate, interpret, solve, and analyze linear equations, functions, graphs, and scatter plots. 2. Interpret and generate transformations. 3. Solve problems involving transversals and parallel lines. 4. Interpret and apply the Pythagorean theorem.

Focus: A Foundational Prerequisite to RTI

Improving mathematics teaching and learning is a significant challenge facing all educators in grades 6–8. For some students, the rapid pace at which topics are covered inhibits their learning. For many, the sheer number of topics per grade level prevents topics from being mastered to the depth and complexity required for even basic retention of knowledge. Mastery of procedures of the overwhelming number of topics is eluding too many, and yet, even mastery of procedures is insufficient. Students must possess mastery of the concepts, procedures, and applications of mathematics topics to succeed as next-generation citizens. RTI is best understood as practices and processes that impact, and are impacted by, virtually everything we do in education (Buffum et al., 2009, 2010, 2012). We leverage the knowledge, skills, and attributes of all members of a learning organization to positively impact all students in a proactive effort to ensure that students receive the supports they need as soon as they show signs of struggle with academic or behavioral expectations. RTI rests on the premise that all students can learn and achieve at high levels and that educators will provide the necessary supports and time needed to ensure this happens. This starts with a culture of belief in every single student and rests on our ability to guide every student to high levels of learning. It also requires that we focus content. The foundation of success for all students rests on the culture of the school and the focus of the students and staff.

This chapter built the case for balancing mathematics content and instruction in all tiers; integrating concepts, procedures, and applications; and favoring depth over breadth. We also suggested a process for prioritizing standards and offered concepts that may emerge as priorities at each grade level.

In the next chapter, we will provide examples of conceptual units at each grade level, including the assessments that can measure student learning of prioritized content. Our objective is not to write a textbook; however, clearly defining and measuring mastery of the content within units of instruction is fundamental to RTI and to all students learning at high levels. In fact, the process of designing units of instruction is the fundamental key to providing differentiated and tiered supports.

Chapter 2

Designing Conceptual Units in Mathematics

This chapter is about *what* we want students to learn and how we can determine whether they have learned it (DuFour et al., 2011). Subsequent chapters describe strategies and pedagogies—the *how*. In this chapter, we will expand beyond the initial prioritization of mathematics content that we shared in the previous chapter to detail the deeper understandings that all middle school students must possess. Deeper understandings are most likely when we have designed rich, focused, and conceptually cohesive units of instruction. After revisiting the prioritization process, briefly describing how to build a unit of instruction, and describing the unwrapping and unpacking of standards, we will begin with the end in mind, with common assessments. Common assessments are an absolutely vital part of the unit design process and represent a key decision-making moment within an RTI process; they establish the target toward which students and staff are working and provide the evidence that can be used to extend learning for some and intervene with others. In the first part of this chapter, we build the case for common assessments and describe how to craft them. In the second half of the chapter, we provide examples of units for grades 6–8.

The Stages of Conceptual Unit Design

When designing units of instruction, our goal must be mastery, not coverage—depth of understanding, not breadth of topics addressed—and we will describe just such a process for building units of instruction. First, how should units be organized and sequenced? We recommend that units be coherently organized, both horizontally within a grade level and vertically between grade levels. Since a sense of number is the basis for all mathematics, we recommend that standards and learning targets that build students' sense of number be frontloaded within a grade level's instructional year. Within grade levels, we recommend that topics such as addition and subtraction, graphing, and algebra be included within individual units and the units that are adjacent to them to allow for more continuity of teaching and learning and greater depth of study. Between grade levels, coherence of topics allows for collaboration between grade levels and Tier 2 intervention, which is particularly relevant in smaller schools where there are one or two teachers per grade level.

Individual units of instruction are best organized around concepts. Conceptual units allow classrooms to explore big ideas in a manner that links key mathematical ways of thinking and connects standards and domains. In forming conceptual units, we recommend the following five-step process.

1. Prioritize the standards.

2. Organize the standards into big ideas.

3. Unwrap the standards into discrete learning targets.

4. Collaboratively create common assessments.

5. Create detailed, viable curriculum maps.

Prioritize the Standards

A prerequisite for designing units of instruction is ensuring that the highly prioritized standards and skills are the focus of a grade level's content. As we have been stressing, prioritizing content is an absolute necessity.

We believe that a primary reason for student frustration and failure, and for the inappropriately elevated rates of determinations for special education eligibility, is that we have been teaching too fast and trying to cover too much content. We are adamant that prioritizing teaching and learning fewer, highly prioritized topics do not represent a lowering of expectations; in fact, we believe that our expectations for mathematics learning must be increased and that expectations will be greater when we require greater levels of depth and complexity of understanding with fewer, highly prioritized topics in a given grade level. Robert Marzano (2001) reports that the single most significant school-level factor in predicting student success is a guaranteed, viable curriculum. Given the number of standards in state curricular frameworks, he asserts that we would need a K–22 school system for all students to demonstrate mastery of all standards. The TIMSS and Programme for International Student Assessment (PISA) reports reveal that students from nations similar to the United States who have superior levels of learning are asked to master eighteen topics per year, versus the eighty topics per year typical in U.S. schools (Gonzales et al., 2008; Organisation for Economic Co-operation and Development, 2011). Mike Schmoker (2011) writes that focus is the most essential element missing from our schools.

Common sense dictates that we prioritize and focus. Even if every standard within a given grade level is deemed equally important—and the authors of the Common Core State Standards for mathematics (NGA & CCSSO, 2010c), like the authors of this book and many other practitioners, have not reached this conclusion—we would have to ask, "What if mastery of every standard is simply not possible? What if we were to value problem solving and critical thinking? What if we were to preassess to identify the gaps in prerequisite knowledge? What if we were to value assessment *as* learning?"

Assessment *as* learning is a process that engages our learners as partners; it is a process in which students identify where they are in their learning journey through self-assessment—perhaps through the use of continuums or rubrics—and establishing learning goals. Assessment *as* learning is closely related to John Hattie's notion of students' self-reporting of grades and the power of timely feedback. Hattie (2012) ranks self-reported grades as the most statistically impactful practice schools can engage in, and feedback has long enjoyed a similarly robust and positive research base.

We must exercise the courage to emphasize depth over breadth and mastery over coverage. Students will learn more, retain more, and be able to apply more. Performance levels on unit tests, even tests that assess everything, will be higher than ever.

How do we prioritize standards? As noted in the previous chapter, teams collaboratively complete a process (Ainsworth, 2003a; Reeves, 2004) like the one shown on page 21. While these steps suggest a mechanical process, we find that prioritizing the most critical big ideas for students to master is actually a highly collaborative process that involves compromise, as teams strive to achieve clarity about the meaning of standards. Once teams embrace the notion that this is a complex process requiring an investment in time, learning, and teamwork, that investment will pay significant dividends.

Organize the Standards Into Big Ideas

After having agreed on a rough prioritization of standards, teams should next organize the prioritized and supporting standards into big ideas and connections. The way standards are written sometimes leaves them open to interpretation. We must build clarity about the meaning of prioritized outcomes and collective consensus on the rigor and format associated with mastery. Big ideas represent a broad set of related standards and skills. We encourage teams to determine the critical topics for which students must demonstrate mastery in a grade level; these are the topics to which a collection of prioritized standards relates.

Teams should then preliminarily map the standards within the school year, projecting the number of ideas that all students can master at depth—in other words, the big ideas that the school will guarantee all students will master because they represent a viable quantity of content. These become the priorities for each grade level.

Unwrap the Standards Into Discrete Learning Targets

Learning targets can serve as objectives that define the scope and focus of a lesson. If a lesson's objective cannot be measured by a single-question exit slip or ticket-out-the-door, then the objective may be too broad. Unwrapping prioritized standards into targets helps teachers and students understand the types of achievement embedded in standards. Teams should experience the collaborative process of unwrapping standards into targets to create a road map for a unit of study. Unwrapping allows teams insight into the relationship between standards, targets, assessment, and instruction so that learning is maximized.

We unwrap standards into discrete targets to inform assessment and instruction. Unwrapping guides teams in their understanding of the building blocks of learning that will lead to mastery of big ideas, so that the instruction can be intelligently and sequentially organized. Teams can also assess more accurately, and more accurate assessments will yield more precise information about the specific strengths and needs of students.

The collaborative process of unwrapping can also inform differentiation. We recommend that teams identify immediate prerequisite skills required to access prioritized content as well as the immediate extension that will result in authentic and worthwhile opportunities to enrich learning for students when they demonstrate mastery.

As we have noted, the standards that we unwrap in this book, and for which we provide sample common assessments, are composites of next-generation standards from the United States and Canada, the National Council of Teachers of Mathematics' recommendations (2000, 2006), and the NGA and CCSSO (2010c) mathematics standards.

When unwrapping standards, we can (Ainsworth, 2003b):

- Underline knowledge targets, or substantive subject content
- Box or circle reasoning targets or thinking skills
- List performances that we must observe and products that students must prepare

For example, figure 2.1 represents a priority for sixth-grade mathematics.

> Use ratio and rate reasoning to solve real-world problems, by reasoning about tables of equivalent ratios, tape diagrams, double number line diagrams, or equations. Make tables of equivalent ratios relating quantities with whole-number measurements, find missing values in the tables, and plot the pairs of values on the coordinate plane. Use tables to compare ratios. Make tables of equivalent ratios.

Figure 2.1: A priority standard for sixth-grade mathematics.

The process of unwrapping standards is not meant to inappropriately deconstruct the big conceptual ideas represented in next-generation standards, like the CCSS, into overly discrete sets of disconnected skills. We embrace and celebrate the fact that we must authentically connect mathematical ideas and embed content within rich and engaging scenarios. However, we must also understand the standards and be able to sequence and teach skills as we build students' conceptual understanding. It's about balance. Unwrapping isn't the end; it's a means to the end.

Collaboratively Create Common Assessments

A next step when designing units of instruction is to collaboratively create common assessments that will be used to measure our success as teachers, determine which students need support with which targets, and inform our current and future practices. As author, researcher, and expert on educational reform Michael Fullan (2005) notes:

> Assessment for learning . . . when done well . . . is one of the most powerful, high-leverage strategies for improving student learning that we know of. Educators collectively at the district and school levels become more skilled and focused at assessing, disaggregating, and using student achievement as a tool for ongoing improvement. (p. 71)

Dylan Wiliam and Marnie Thompson (2007) identify formative assessments as transformative processes:

> Effective use of formative assessment, developed through teacher learning communities, promises not only the largest potential gains in student achievement but also a process for affordable teacher professional development. (p. 36)

Common assessments are proven tools that can dramatically improve student learning. Studies have demonstrated that they rival one-on-one tutoring in their effectiveness and that the use of assessments benefits low-achieving students in particular (Stiggins, 2007). Well-designed assessments allow teams to measure and diagnose learning on a by-the-target basis and require students to solve for, explain, and justify solutions.

The overwhelmingly dominant form of assessment in schools is selected response, and yet selected-response tests are valid and accurate assessments only for knowledge, which is just a fraction of the learning targets for which we want students to demonstrate mastery and therefore measure. There are four primary types of learning targets (Stiggins, 2007).

1. **Knowledge:** Involves mastery of content that represents knowing and understanding. The most efficient method for assessing here is selected response—matching, true or false, and multiple choice.

2. **Reasoning:** Involves using knowledge to draw conclusions and solve problems. The most appropriate method for assessing is extended response that requires a description of thinking.

3. **Performance:** Involves completing a task in which process is as important as the product. Example methods of assessment include playing a musical instrument, orally defending a position, reading aloud, speaking in a second language, and using a psychomotor skill.

4. **Product:** Involves creating tangible products that represent mastery. Example methods of assessment include written work, science fair models, and art products.

If we value knowledge, reasoning, performance, and product, then we must design, administer, and analyze assessments that measure these items. Importantly, if we are going to assess in these ways, our instruction must prepare students for these tasks.

Common assessments are the key lever to RTI within Tiers 1 and 2. In backward planning from common assessments, teams ensure that their instruction prepares students for the rigor and format required of students to achieve mastery. In sharing expectations represented in the common assessments, teachers provide students with clear targets. In analyzing student work when completing common assessments, teams gauge their success at ensuring high levels of learning for all and can learn from one another when some teachers experience relatively greater levels of success. Teams can also identify which students need more time to master which prioritized standards and can collaboratively determine the strategies that are needed to support them.

Timely evidence gathering and responses to student needs result in the most successful interventions. We recommend that teams and teachers commit to administering midlesson checks for understanding and end-of-lesson exit slips to determine class success at mastering daily targets. Midlesson checks for understanding can be accomplished within student notebooks, on scrap paper, or on personal whiteboards. They allow the teachers and students to gauge progress midway through a lesson when making the crucial decision as to whether students are ready to practice at greater levels of independence or require more time and an alternative explanation. End-of-lesson exit slips allow teachers and teams to measure the success of lessons and provide the evidence to inform the next day's instruction and any necessary reteaching.

Common assessments administered during and at the conclusion of units can answer five critical questions when collaboratively analyzed by teacher teams.

1. Who has not yet learned the prioritized standards within the unit?
2. For what specific skills must we devote more time and employ different assessment and instructional approaches?
3. Which teacher on the team has had more relative success with these skills?
4. If there is not a teacher on the team who has had more success, are there teachers within the broader school system whom we can contact?
5. What external resources, strategies, or professional development can we access?

Without the common assessments teacher teams administer and analyze, we lack the timely and specific evidence we need to inform these targeted supports. We cannot overstate this critically important point: while assessment *for* learning and assessment *as* learning are powerful levers for dramatically improving student achievement, they will not occur without greater focus. While preassessments, universal screening, and common formative assessments have been shown in the research to improve and deepen all students' learning, they take time. We will not have the time if we continue to race through the curriculum; we must focus on student mastery of the most highly prioritized content by collaboratively planning for a guaranteed, viable curriculum.

Preassessments and Universal Screening

A vital category of common assessments that can significantly inform and motivate learning is preassessments and universal screening. Preassessments administered before the beginning of units can be similar or identical to the common assessments given at the conclusion of units and should help schools predict which students lack immediate prerequisite skills. This allows us to proactively preteach students before and at the beginning of new units. Some districts and schools may already require that teachers administer preassessments, but staff members may not be certain *why* they give these tests. It's also possible that staff members know why they give these tests, but schools do not provide time for the collaborative analysis of the results. (Screening and preassessments are both preventive, predictive assessments.)

As we increasingly differentiate for individuals, groups, and entire classes, clarity regarding where the consolidation of some content will provide additional instructional time for other content is incredibly valuable. The ability to compact instruction will allow teachers and students to use precious time more efficiently. If schools build time to assess and preteach into schedules, giving teachers advanced knowledge of which students require preteaching support and the prerequisite skills of which they lack understanding, educators can provide proactive, preventive supports.

Scoring Guides

The use of common assessments that more authentically measure student learning and more accurately diagnose student strengths and needs requires that we use more open-ended items. This will require the use of scoring guides that educators and students can use to commonly, consistently, and accurately assess student learning. Collaboratively developed scoring guides ensure that education teams have a common understanding of proficiency. Accurate use of scoring guides requires that teams work toward greater consistency in the assessment practice. Anchor solutions or exemplars—examples of student work that represent different levels of mastery—will also need to be collected. Figure 2.2 shows a sample scoring guide. Working from right to left, teachers first evaluate a student's answer: is the answer correct, partially correct, or irrational? Next, teachers evaluate a student's solution: does the evidence represent a complete and coherent process that illustrates the student's thinking, is the process partially complete and coherent, or is evidence of thinking absent? Lastly, teachers examine a student's understanding: to what extent do the descriptions of why steps were taken or the justifications of the solution represent an understanding of why and how the mathematics works?

Student Self-Assessment

As we have noted, Hattie (2009) identifies self-reported grades as the most powerful strategy in which students can engage. Self-reported grades require that students self-assess their own work and set goals for the next steps they must take and the additional support they need to further their learning. This is how the process works: After students complete a common assessment, they are provided with a scoring guide similar to the one in figure 2.2. They are also introduced to an anchor solution or exemplar so that they can compare their work to a solution that represents mastery. Once students have been guided through the use of the scoring guide and the process of self-assessing, they can set goals, identify needs, and take more

Question	Student's Understanding	Student's Solution	Student's Answer
	2 Understands the problem	2 Develops a plan that could lead to the correct answer	2 Offers a correct solution
	1 Misinterprets parts of the problem	1 Gives a partially correct procedure	1 Makes a copying or computational error; gives a partial answer, provides no answer statement, or labels answer incorrectly
	0 Makes no attempt or misinterprets problem	0 Makes no attempt or provides an incorrect plan	0 Provides no answer or an illogical answer

Figure 2.2: Sample scoring guide.

responsibility for their own learning. When students self-assess, they begin to assume greater responsibility for their own learning. Through self-assessment, students become independent learners, capable of setting their own goals. When students evaluate their own evidence of learning against an exemplar, they are encouraged to identify and analyze their errors, thereby extending the learning process into the analyses of the assessment itself.

The Importance of Data

The evidence of learning is the engine that drives RTI and that will propel all students toward higher levels of learning. Schools, teams, teachers, and students must be hungry for targeted information that will help identify where we are in the learning journey and what next steps are required. Assessments have gained a negative reputation; yet without the data we derive from assessments—formal and informal; graded and nongraded; before instruction, during instruction, and after instruction—the evidence necessary to determine current levels of readiness and inform future learning will not be available.

Create Detailed, Viable Curriculum Maps

Lastly, teams will revisit the preliminary maps they constructed after prioritizing standards and chunking them into big ideas and turn them into more detailed maps. They will ensure that the prioritized and unwrapped standards that they have designed assessments for can be viably mastered by all students to the requisite level of depth and complexity, taking into account time for:

- Building students' conceptual understanding

- Building students' procedural understanding

- Authentic problem-solving and application tasks

- Preteaching and previewing prerequisite skills when evidence determines the need

- Reteaching prioritized skills when evidence determines the need

- Enriching understanding of prioritized skills when students are ready

- Using assessment *as* learning and involving students intimately in the learning process

The process of (1) prioritizing standards, (2) chunking them into big ideas, (3) unwrapping standards into learning targets, (4) collaboratively designing common assessments, and (5) mapping a detailed, viable curriculum can be applied to any curricular area. Such a process is an absolute prerequisite to building deeper levels of understanding for all students in mathematics.

Sample Conceptual Unit for Seventh Grade

In the remainder of this chapter, we apply the processes that we have described to seventh-grade mathematics. Rational numbers, including integers, proportional reasoning, and expressions and equations are critical to seventh-grade mathematics. Visit **go.solution-tree.com/mathematics** for examples for sixth and eighth grade. We reiterate: the work that follows is an example only, not the answer. We suggest big ideas, prioritized standards, unwrapped targets, and mapped units as an option, not as the optimal solution. The process in which teams engage to complete this vital professional work is as important, or more important, than the lists that teams produce. Teams should collaboratively engage in this important work. We hope that the ideas and resources within this chapter help teachers begin the work of crafting rigorous units of instruction.

Prioritize the Standards

Prioritized standards for interpreting rates and ratios are listed in table 2.1. This content provides the foundation for more complex algebraic work with rational numbers.

Table 2.1: Interpreting Rates and Ratios Standards for Grade 7

Analyze proportional relationships.	Compute unit rates associated with ratios of fractions, lengths, areas, and other quantities measured in like or different units.
Use proportional relationships to solve real-world problems.	Recognize and represent proportional relationships.
Identify the constant of proportionality or unit rate in tables, graphs, equations, diagrams, and verbal descriptions.	Decide whether two quantities are in a proportional relationship.
Use proportional relationships to solve multistep percent problems.	Represent proportional relationships by equations.
Explain proportional relationships on a coordinate plane by computing the unit rate formed between the coordinates of ordered pairs.	

Source: Adapted from NGA & CCSSO, 2010c.

Table 2.2 shows prioritized standards for decimals, fractions, and percents, which allow for the equivalent representation of numbers.

Table 2.2: Decimals, Fractions, and Percents Standards for Grade 7

Convert a fraction to a decimal and compare.	Convert a fraction to a percent and compare.
Convert a decimal to a rational number and compare.	Convert a percent to a fraction and compare.
Convert a decimal to a percent and compare.	

Source: Adapted from NGA & CCSSO, 2010c.

Table 2.3 (page 38) shows prioritized standards for rational number computation, which allows for the proficient and fluid problem solving of all mathematics problems.

Table 2.3: Rational Number Computation Standards for Grade 7

Understand addition and subtraction of integers as the sum or differences of the numbers' absolute values.	Understand subtraction of integers as adding the additive inverse.
Show that the distance between two rational numbers on the number line is the absolute value of their difference.	Interpret sums of rational numbers by describing real-world contexts.
Add, subtract, multiply, and divide all rational numbers.	Interpret differences of rational numbers by describing real-world contexts.
Understand the rules of signs when multiplying and dividing integers.	Solve real-world and mathematical problems involving the four operations with rational numbers.

Source: Adapted from NGA & CCSSO, 2010c.

Table 2.4 shows the prioritized standards for multistep equations and inequalities. This is foundational to work with multivariable linear equations.

Table 2.4: Multistep Equations and Inequalities Standards for Grade 7

Generate equivalent expressions.	Add, subtract, factor, and expand expressions with rational-number coefficients.
Rewrite an expression to better or more clearly understand a situation.	Solve real-world problems, expressions, and equations.
Solve multistep real-world problems with all positive and negative rational numbers.	Use variables to construct simple equations and inequalities.
Assess the reasonableness of answers using estimation.	Solve word problems by writing and solving multistep equations.
Solve word problems by writing and solving multistep inequalities.	

Source: Adapted from NGA & CCSSO, 2010c.

Table 2.5 lists the prioritized standards for computing with circles and prisms, which builds directly on knowledge of basic computation and measurement.

Table 2.5: Computing With Circles and Prisms Standards for Grade 7

Know and use the formulas for the area and circumference of a circle.	Solve real-world problems involving angle measure and area.
Describe the relationship between the circumference and area of a circle.	Solve real-world problems involving the volume and surface area of three-dimensional objects composed of triangles, quadrilaterals, polygons, cubes, and right prisms.

Source: Adapted from NGA & CCSSO, 2010c.

Organize the Standards Into Big Ideas

Big ideas include:

1. Proportions and percents

2. Decimals, fractions, and percents

3. Rational number computation

4. Multistep equations and inequalities

5. Circles and prisms

A rough mapping of the big ideas of seventh grade (table 2.6) provides students with nine weeks to master proportions and percents; six weeks to manipulate decimals, fractions, and percents; seven weeks to compute with all rational numbers; eight weeks to solve multistep equations and inequalities; and six weeks to compute the area and circumference of circles and the surface area and volume of prisms.

Table 2.6: Preliminary Mapping of Prioritized Big Ideas

45 Days	30 Days	35 Days	40 Days	30 Days
Proportions and percents	Decimals, fractions, and percents	Rational number computation	Multistep equations and inequalities	Circles and prisms

Work with proportions, equations, and computational measurement should be the focus of seventh-grade mathematics. The big ideas could more fully be described by the following statements.

- Real-world problems, including those involving percents, can be solved by setting up and solving proportions.

- All rational numbers can be expressed in three equivalent forms—fractions, decimals, and percents.

- Using knowledge of place value and decomposition, all rational numbers can be added, subtracted, multiplied, and divided.

- Real-world situations can be represented by expressions and equations, and expressions and equations can be logically evaluated and solved.

- The circumference and area of circles are based in the wonderful ratio called π or pi; the surface area and volume of prisms simply require knowledge of the area of simple figures.

Unwrap the Standards Into Discrete Learning Targets

Teams' collaborative unwrapping of prioritized standards enriches and informs assessment and instruction. Tables 2.7–2.11 (pages 40–42) show sample unwrapped targets of seventh-grade prioritized standards.

Table 2.7: Unwrapped Proportions and Percents Targets for Grade 7

I can . . .	
Compute unit rates given fractional terms	Calculate and explain the factor by which terms are multiplied (called the *constant of proportionality* or *unit rate*) in proportional relationships through: • Tables • Graphs (the slope) • Equations (*m*) • Diagrams • Written descriptions
Prove that two terms are proportional by cross-multiplying (a shortcut for finding a common denominator)	Write equations that represent proportional relationships
Prove that two terms are proportional by creating a table of terms	Interpret the meaning of coordinates in graphs that represent proportional relationships
Use proportional reasoning to solve problems involving: • Simple interest $I = p \times r \times t$ • Taxes, markups, tips, gratuities, commissions, and fees $\dfrac{tax}{cost} = \dfrac{\%}{100}$ • Discounts $\dfrac{discount}{cost} = \dfrac{\%}{100}$ • Percent increase or decrease $\dfrac{change\ in\ price}{original\ price} = \dfrac{\%}{100}$ • Percent error $\dfrac{measured\ value - calculated\ value}{calculated\ value} = \dfrac{\%}{100}$	Prove that two terms are proportional by plotting in the Cartesian plane

Table 2.8: Unwrapped Decimals, Fractions, and Percents Targets for Grade 7

I can . . .	
Divide to convert fractions to decimals	Know that decimals will terminate or repeat
Use place value to convert from decimals to fractions	Use place value to convert from decimals to percents
Divide to convert from percents to fractions	Use proportions to convert from fractions to percents
Convert from mixed numbers to improper fractions	Use place value to convert from percents to decimals

Table 2.9: Unwrapped Rational Number Computation Targets for Grade 7

I can . . .	
Add integers	Subtract integers
Add rational numbers, including sums in which addends are negative	Subtract rational numbers, including sums in which subtrahends and minuends are negative
Use number lines to visualize the addition of integers and negative rational numbers	Use number lines to visualize the subtraction of integers and negative rational numbers
Create real-world scenarios involving the addition of integers and negative rational numbers	Create real-world scenarios involving the subtraction of integers and negative rational numbers
Add rational numbers	Know and apply the rules for multiplying signed rational numbers
Subtract rational numbers	Multiply signed rational numbers
Know and apply the rules for dividing signed rational numbers	Create real-world scenarios involving the multiplication of signed rational numbers
Divide signed rational numbers	Multiply rational numbers
Create real-world scenarios involving the multiplication of signed rational numbers	Divide rational numbers

Table 2.10: Unwrapped Multistep Equations and Inequalities Targets for Grade 7

I can . . .	
Simplify expressions by combining like terms	Write and solve equations representing real-world problems in the form $px + q = r$, both arithmetically and algebraically
Expand expressions by factoring	Analyze both arithmetic and algebraic solutions and compare the required steps

continued →

Simplify and expand expressions to clarify the meaning of real-world problems	Write and solve equations representing real-world problems in the form $p(x + q) = r$, both arithmetically and algebraically
Write, solve, and graph (on a number line) inequalities representing real-world problems in the form $px + q > r$	Analyze both arithmetic and algebraic solutions and compare the required steps

Table 2.11: Unwrapped Circles and Prisms Targets for Grade 7

I can . . .	
Solve for the area of a circle using the proper formula	Find the area of polygons that may include triangles and quadrilaterals
Solve for the circumference of a circle using the proper formula	Find the area of composite polygons that may include triangles and quadrilaterals
Compare and analyze the formulas for finding the area and circumference of a circle	Find the volume of right prisms
Find the surface area of right prisms	Find the volume of composites of right prisms
Find the surface area of composites of right prisms	

Supporting standards can contribute to, or extend, student understanding of seventh-grade big ideas (see tables 2.12–2.16).

Table 2.12: Supporting Proportions and Percents Standards for Grade 7

Approximate the probability of an event by collecting data.	Understand that the probability of a compound event is the fraction of outcomes in the sample space for which the compound event occurs.
Predict relative frequencies of an event occurring given a probability.	Design and use a simulation to generate frequencies for compound events.
Solve problems involving scale drawings.	Represent sample spaces for compound events using lists, tables, and tree diagrams.
Explain proportional relationships on a coordinate plane by computing the unit rate formed between the coordinates of ordered pairs.	

Source: Adapted from NGA & CCSSO, 2010c.

Table 2.13: Supporting Decimals, Fractions, and Percents Standards for Grade 7

Understand that the probability of an event is a number between 0 and 1.	Develop a probability model by assigning equal probability to all outcomes.
Develop a probability model by observing frequencies in data generated from a chance process.	

Source: Adapted from NGA & CCSSO, 2010c.

Table 2.14: Supporting Rational Number Computation Standards for Grade 7

Describe situations in which opposite quantities combine to make 0.	Informally assess the degree of visual overlap of two numerical data distributions with similar variabilities.
Use measures of center and measures of variability from random samples to draw inferences.	Apply properties to add and subtract rational numbers.

Source: Adapted from NGA & CCSSO, 2010c.

Table 2.15: Supporting Multistep Equations and Inequalities Standards for Grade 7

Understand that statistics inform a population.	Use data from a random sample to draw inferences.
Understand that generalizations about a population from a sample are valid only if the sample is representative.	Generate multiple samples of the same size to determine variation.
Use facts about supplementary, complementary, vertical, and adjacent angles in a multistep problem to write and solve simple equations for an unknown angle in a figure.	

Source: Adapted from NGA & CCSSO, 2010c.

Table 2.16: Supporting Circles and Prisms Standards for Grade 7

Draw geometric shapes with given conditions.	Describe the two-dimensional figures that result from slicing three-dimensional figures.

Source: Adapted from NGA & CCSSO, 2010c.

Student learning develops along a continuum. Investigating the continuum of learning when unwrapping standards can further inform teams' knowledge of prioritized content. This practice facilitates differentiation of instruction. Figure 2.3 (pages 44–45) is an example of a learning continuum for prioritized seventh-grade standards.

Recognize opposite signs of numbers as indicating locations on opposite sides of 0 on the number line.	Understand p + q as the number located a distance |q| from p in the positive or negative direction depending on whether q is positive or negative.	Know that numbers that are not rational are called irrational and that every rational number has a decimal expansion that repeats or terminates.

An example of a portion of a seventh-grade common assessment follows.

1.

    a. Jorge had $3,400 in his bank account. He spent $1,600 on a used car. How much money does he have left in his account?

    b. Express each of the money amounts using integer signs.

    c. Prove your answer by describing the rules, providing a proof, or using a picture, drawing, or diagram.

2.

    a. Evaluate the expression $-15 + 23 - 8 + 9$

    b. Prove your answer by describing the rules or providing a proof.

3.

    a. A friend made the following mistake when solving this problem. Explain and correct your friend's error.

       $2\,^{5}/_{12} + 3\,^{7}/_{12} = 5\,^{12}/_{24}$

    b. Complete the problem correctly.

4.

    a. Dylan went to the hardware store and bought 2 yellow ropes of the same length. The total length of the ropes was 14.9 meters. How long was each rope?

    b. Use a picture, drawing, model, or diagram to prove your answer.

5.

    a. Mariceli's bucket holds $^{1}/_{2}$ of a cup of sand. Mariceli is filling the bucket with a scoop that holds $^{1}/_{6}$ of a cup. How many scoops of sand will it take for Mariceli to fill the bucket?

    b. Simplify your answer and write it as a proper fraction or as a whole or mixed number.

    c. Use a picture, drawing, model, or diagram to prove your answer.

6.

    a. Divide $8 \div \frac{1}{5} =$

    b. Use a picture, drawing, model, or diagram to prove your answer.

7.

    a. A case of canned fruit cocktail costs \$4.75. Eduardo bought $2\frac{1}{2}$ cases. How much did Eduardo spend?

    b. Use a picture, drawing, model, or diagram to prove your answer.

8.

    a. A friend made the following mistakes when solving this problem. Explain and correct your friend's errors.

       $-2\frac{2}{3} \times 3\frac{1}{3} = 5\frac{2}{9}$

    b. Complete the problem correctly.

9.

    a. Write 0.3 as a fraction.

    b. Use a picture, drawing, model, or diagram to prove your answer.

Figure 2.3: Sample learning continuum for prioritized seventh-grade standards.

A sample knowledge package (Ma, 1999) for seventh grade is provided in figure 2.4. Knowledge packages guide teams as they further unwrap standards and brainstorm concepts that connect to, and connect from, big ideas.

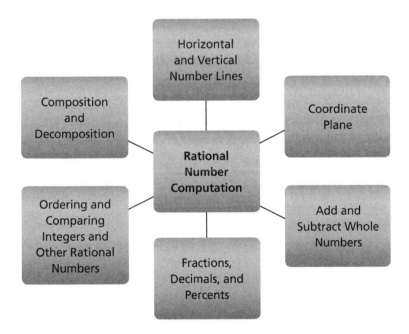

Figure 2.4: Sample knowledge package for seventh grade.

Create Detailed, Viable Curriculum Maps

Teacher teams next expand their preliminary mapping of prioritized big ideas to organize standards into cohesive and sequential coherence units. Table 2.17 includes a general mapping of prioritized seventh-grade standards.

Seventh-grade mathematics should focus on developing mastery of operations with all rational numbers, proportional reasoning that includes percents, and multistep expressions and equations in one variable. Eighth grade builds directly on this foundation, exploring linear equations in two variables.

Deeper Understanding of RTI for All

Response to intervention is for all students. We must strive to ensure that all students are more deeply responding to instruction, and this is particularly true for mathematics. It is no longer adequate for us to simply require students to arrive at the correct answer and to show their work in an organized and clear manner, as this may allow them to camouflage a lack of conceptual understanding; students "showing their work" may maintain an overdependence on rote computation skills. Students must also demonstrate an understanding of their solutions and be able to prepare visual representations and justify them with written explanations. Creating visual representations and providing justifications help students acquire a deeper understanding of mathematical concepts and problem-solving skills. For students to achieve these goals, we must prepare them—and ourselves. Assessment, and other ways of gathering evidence, must also be transformed. Assessment tasks must be more authentic and open ended, and teachers should utilize scoring guides or rubrics to collaboratively analyze student understanding. A shift from a near-universal reliance on selected-response assessments will allow us to finally and completely move from assessment *of* learning to assessment *for* learning. Educators will then have the evidence and information to determine where student understanding is strong and where a lack of understanding requires more support. Assessment will more accurately inform learning and will be used for future learning. We propose goals that are even loftier. We know that student self-assessment is among the most statistically impactful practices in which we can engage (Hattie, 2009). Building time into instructional units, preparing students to analyze their own work, and building on students' recognition of strengths and continued needs will transform testing into assessment *as* learning.

Table 2.17: General Mapping of Prioritized Grade 7 Standards

Unit 1 Decimals, Fractions, and Percents 30 Days	Unit 2 Proportions and Percents 30 Days	Unit 3 Integer Computation 30 Days	Unit 4 Computation With all Rational Numbers 30 Days	Unit 5 Multistep Equations and Inequalities 30 Days	Unit 6 Circles and Prisms 30 Days
Convert a fraction to a decimal and compare.	Analyze proportional relationships.	Understand addition and subtraction of integers as the sum or differences of the numbers' absolute values.	Add, subtract, multiply, and divide all rational numbers.	Generate equivalent expressions.	Know and use the formulas for the area and circumference of a circle.
Convert a decimal to a rational number and compare.	Use proportional relationships to solve real-world problems.	Show that the distance between two rational numbers on the number line is the absolute value of their difference.	Solve real-world and mathematical problems involving the four operations with rational numbers.	Rewrite an expression to better or more clearly understand a situation.	Describe the relationship between the circumference and area of a circle.
Convert a decimal to a percent and compare.	Identify the constant of proportionality or unit rate in tables, graphs, equations, diagrams, and verbal descriptions.			Solve multistep real-world problems with all positive and negative rational numbers.	Solve real-world problems involving the volume and surface area of three-dimensional objects composed of triangles, quadrilaterals, polygons, cubes, and right prisms.
Convert a fraction to a percent and compare.	Use proportional relationships to solve multistep percent problems.	Understand the rules of signs when multiplying and dividing integers.		Assess the reasonableness of answers using estimation.	Solve real-world problems involving angle measure and area.
Convert a percent to a fraction and compare.	Compute unit rates associated with ratios of fractions, lengths, areas, and other quantities measured in like or different units.	Understand subtraction of integers as adding the additive inverse.		Solve word problems by writing and solving multistep equations.	
Convert a fraction to a percent and compare.	Recognize and represent proportional relationships.	Interpret sums of rational numbers by describing real-world contexts.		Add, subtract, factor, and expand expressions with rational-number coefficients.	
	Decide whether two quantities are in a proportional relationship.	Interpret differences of rational numbers by describing real-world contexts.		Solve real-world problems, expressions, and equations.	
	Represent proportional relationships by equations.			Use variables to construct simple equations and inequalities.	
				Solve word problems by writing and solving multistep inequalities.	

Source: Adapted from NGA & CCSSO, 2010c.

To be successful as 21st century learners and to graduate ready for college and a skilled career, students must (NGA & CCSSO, 2010d):

1. Make sense of problems and persevere in solving them.

2. Reason abstractly and quantitatively.

3. Construct viable arguments and critique the reasoning of others.

4. Model with mathematics.

5. Use appropriate tools strategically.

6. Attend to precision.

7. Look for and make use of structure.

8. Look for and express regularity in repeated reasoning. (pp. 6–8)

These eight habits of mind are the Standards for Mathematical Practice identified by the CCSS (NGA & CCSSO, 2010d). We must prepare students for these lofty goals, provide them with opportunities to practice tasks that represent them, and offer frequent and specific feedback to students as they work toward them. We must also have the time to accomplish these tasks. Creating the time requires a focus on depth over breadth and the design of conceptually sound and connected units of instruction.

Looking ahead, chapter 3 will explore what conceptual understandings are required to teach and learn the prioritized standards and will examine the afore-mentioned habits of mind. Chapter 4 will move from what mathematics content is critical for students to learn to what instructional practices will ensure student learning and mastery of these prioritized mathematics concepts. We will detail the content knowledge that teachers must possess to ensure that students master the most highly prioritized middle school concepts and the strategies and approaches that will provide all students with the supports that they need to learn at high levels. Chapter 5 will provide concrete ideas that schools can use to intervene when students require supplemental supports to master this critical content.

Chapter 3

Instructional Practices for Application-Based Mathematical Learning

After educators have identified essential outcomes or big ideas and designed conceptual units that promote the interrelated nature of mathematics, they must focus their attention on the instructional practices that promote authentic, application-based learning. *Application-based learning* is an instructional mindset that shifts the focus from ensuring students can use mathematics to compute answers to ensuring conceptual understanding and procedural competency *applied* to real-world problems. This chapter will provide guidance related to the three interrelated elements of highly effective Tier 1 instruction: (1) teachers' knowledge of the mathematics, (2) the integration of mathematical practices (habits of the mind), and (3) effective instructional practices. In addition, the chapter will provide a sample structure to organize mathematics instruction that focuses on application-based mathematics learning.

Develop Instruction

The complexity of mathematical learning is such that students must master concepts and skills with an understanding that is fluid enough to allow them to apply these concepts and skills in a variety of contexts. In addition, students must be able to communicate and model mathematics, and explain and justify their solutions. Students must also master and apply related academic language and mathematical tools. For teachers to design and implement instruction that ensures such rich learning, they must be able to weave evidence-based instructional strategies with mathematical practices and apply those elements strategically within engaging, real-world contexts. Effective instruction includes those instructional decisions that positively impact student learning and engagement. The NMAP (2008) identifies these practices as follows:

- Maintenance of the balance between student-centered and teacher-directed instruction

- Explicit instruction for students having mathematics difficulties

- Simultaneous development of conceptual understanding, procedural competency, and application of skills in problem-solving contexts

- Sufficient practice to develop automatic recall of basic facts and fluency with procedures

- Social and emotional involvement of peers and cooperative learning experiences

- Emphasis on the importance of effort (effort has greater impact on mathematical performance than inherent talent)

- Ongoing use of formative assessment

These practices, in combination with the recognition that such learning requires sufficient time, can inform our instructional designs. As we consider the needed shifts in instructional practices, we recognize the need to focus instruction on a viable curriculum. And yet, simply shifting to a new set of learning standards and greater prioritization of these standards (while critical) will not improve student learning if effective instructional practices are not promoted. We also recognize that simply providing a menu of instructional activities without focused and rigorously designed units of instruction will not significantly improve student learning. The current challenge for mathematics teachers as we continue to make significant shifts in instruction and learning is multifaceted.

First, traditional, competitive, noncollaborative instructional models have tainted the experience of many mathematics learners. Mathematics lessons have been procedurally based, with an emphasis on rote practice, often at the exclusion of opportunities for developing conceptual understanding and the authentic application of mathematics within problem-solving situations. The traditional instructional model did little to promote confident, fluid mathematical thinking or a genuine love of mathematics for many of us (Ball, 2005). Without improvements in teacher confidence, new instructional practices that promote these areas will likely feel unnatural and difficult to implement. These realities may lead to perceived or actual resistance to changing instructional practices. Teachers have a desire to provide positive, effective learning experiences for their learners. They're not resistant to changing instructional practices, but they need to be empowered to do it successfully. Kerry Patterson, an innovator in change and leadership, and his colleagues (2008) explain that for educators to embrace changing practices, two questions must be positively answered: (1) Is it worth it? and (2) Can I do it?

Our answer to each question is a resounding yes! Improving mathematics instruction and achievement is critical for our students. We can do so if we provide our teachers with the opportunity to learn, explore, and experience new mathematical practices that promote student engagement and understanding. Author and professor Deborah Ball and colleagues (Ball, Bass, & Hill, 2004) find that teachers must be able to both perform the mathematics they are teaching *and* "think from the learner's perspective and . . . consider what it takes to understand a mathematical idea for someone seeing it for the first time" (p. 21).

Collaborative Development

The interweaving of content knowledge (understanding the mathematics), mathematical practices, and evidence-based instructional practices creates dynamic, rich, responsive learning experiences that contribute to significant improvements in student achievement. Collaborative teacher teams and PLC at Work™ cultures greatly enhance this work and should intentionally focus on these three elements. As educators clearly define prioritized learning outcomes and develop common assessments, they are strengthening their own conceptual understanding of mathematics and benefiting from a variety of other perspectives. As teacher teams clearly define desired outcomes, they are proactively identifying prerequisite skills and common misconceptions that may interfere with student learning. Teacher teams should also examine key mathematical processes to identify which practices will have the most impact on the prioritized learning outcome being discussed. While we recognize that each teacher must adapt instructional experiences based on learner needs, collaborative planning of the instructional process can provide educators with a framework for instruction. The effectiveness of teacher collaboration is greatly enhanced within the context of a targeted, focused task. Table 3.1 (page 52) identifies suggested focuses for collaboratively developing high-impact mathematics instruction.

Table 3.1: Collaborative Teacher Teams—Developing Highly Effective Instruction

Element of Highly Effective Instruction	Teacher Team Tasks
Mathematical Content Knowledge	1. Clearly define the prioritized learning outcomes in teacher-friendly and student-friendly language. 2. Identify, define, and clarify the mathematical vocabulary and language that will be used. 3. Identify what prerequisite skills may interfere with student learning. Discuss how to identify if students have deficits in these areas.
Mathematical Practices	1. Review key mathematical practices and processes. 2. Identify practices that most authentically connect to prioritized learning outcomes. 3. Plan for intentional application of practices.
Instructional Practices and Strategies	1. Frame instructional practices within phases of instruction. 2. Recognize required adaptations for student needs (whole class, small group, and individual).

Teacher Knowledge

For example, Travis Port and the other grade 7 teachers at Myers Middle School identify solving real-world proportional problems as a prioritized learning outcome for the first term. As they begin to collectively plan instruction, they identify three instructional elements that they can collectively address and support.

1. **Educators' depth of knowledge related to proportional reasoning, multiplicative relationships, rate, ratio, and percent:** Teachers must know and understand the mathematics they teach in order to effectively plan instruction and answer the following questions. What are the prerequisite skills my students must have to engage in this learning? How will I extend learning for students who have mastered this? What common misconceptions might students experience? In what different ways can this mathematics be represented? What mathematical tools should be available for student learning?

2. **Educators' knowledge and understanding of the mathematical practices that can be embedded in instruction:** Identifying patterns and repeated reasoning, and identifying structures and using them in thinking are two of the mathematical practices that connect authentically to using proportions to solve real-world problems. As students engage in related

learning tasks, it is important to pose questions such as, What different ratios can be used to explain this situation? and What is the relationship between equivalent ratios and equivalent fractions? Students should also identify and explain various ways to represent the mathematics such as graphs, rate tables, or concrete manipulatives. Mathematical practices will be discussed more thoroughly in the following section.

3. **Educators' use of evidence-based instructional practices that match the learning outcomes:** How will instruction be differentiated to meet student needs? What vocabulary should be taught? Should students explore representing proportional problems with manipulatives, by drawing representations, or by using both methods? How can we promote both individual ownership and collaboration in our learning experiences? What metacognitive, formative strategies will motivate or engage which students? What instructional strategies will best support students mastering this concept?

These three intertwined elements are the basis of highly effective, application-based mathematics instruction that will promote mastery of essential learning outcomes. Recognizing the impact that these elements have on student learning and ensuring that instruction addresses all three areas allow the team to provide effective and meaningful Tier 1 instruction for students. Figure 3.1 features a sample collaborative Tier 1 instruction planning grid for teacher teams.

Essential learning outcome (in student-friendly language): We can use rates and ratios to solve real-world problems. These problems include finding scale, rates, and ratios. Our learning will help us understand fractions, percents, and decimals.	
Probing Questions	**Team Planning Notes**
What are the prerequisite skills our students must have to engage in this learning?	• *Multiplication* • *Comparing quantities* • *Repeated reasoning*
If students have deficits in the prerequisite skills, how can we scaffold their learning?	• *Small-group preteach—first two days of conceptual unit* • *Calculators* • *Concrete modeling*

Figure 3.1: Myers Middle School seventh-grade collaborative teacher team planning grid (Tier 1 instruction). continued →

What are the big ideas associated with this outcome?	• *Proportions involve multiplicative relationships.* • *Rates, ratios, percents, fractions, and decimals are comparisons.* • *Ratios compare equal units (1 cup / 3 cups).* • *Rates compare different units.* • *Rates and ratios can compare part-to-part or part-to-whole.* • *Ratios can be represented as a fraction (Small, 2009b).*
What common misconceptions might students experience?	• *Comparing units* • *Multiplicative relationships* • *Part-to-part and part-to-whole*
How will we extend learning for students who have mastered this?	*Challenge students to represent proportional reasoning in more than one way—part-to-part and part-to-whole; whole-to-part and reverse part-to-part.*
How can this mathematics be presented concretely?	*Manipulatives and virtual manipulatives*
How can this mathematics be represented pictorially?	*Drawings, tables, and digital pictures*
What are the abstract ways to represent this math?	*Fractions, ratios, and graphs*
What mathematical practices (habits of the mind) connect to this outcome?	• *Make sense of problems/persevere.** • *Reason abstractly/quantitatively.** • *Construct viable argument/critique reasoning.* • *Model.* • *Use appropriate tools.* • *Attend to precision.*

	• *Make use of structure.* • *Express regularity in repeated reasoning.** ** Focus practices for this unit— student work samples to be shared*
What mathematical tools should be available for student learning?	*All manipulatives, computers, laptops, iPods, and calculators*
How could instruction be differentiated to meet student needs?	*Preteach, small group, and mini-lessons*
What vocabulary should be taught?	*Ratio, rate, proportions, part-to-part ratio, part-to-whole ratios, and units*
What collaborative experiences should be integrated into conceptual unit design?	*Use homogeneous grouping for this conceptual unit.*
What individual experiences should we integrate into unit design?	• *Individual think time* • *Practice activities including individual component, ticket-out-the-door with student name, and mini-quiz*
What metacognitive, formative assessment practices should be utilized?	*Observation, ticket-out-the-door, individual tweet or mini-blog entry, student think-aloud, and journal entry*
What instructional strategies will best support students' mastery of this concept?	*Journals and C-P-A*

Integration of Practices to Promote Application-Based Mathematics Learning

Given the complexity of mathematics instruction, this chapter examines the mathematical practices that must be integrated into learning experiences and provides concrete, evidence-based instructional practices that promote deep, authentic learning. Interweaving these facets of middle school mathematics will serve as the catalyst for ongoing instructional improvements in middle school classrooms. These instructional practices collectively represent an instructional process that can best be described as *application based*, our term for mathematics instruction that focuses on the end in mind. Our goal, through all mathematics instruction, is that our students acquire the authentic mathematics skills that prepare them to be what the introduction describes as future ready (Rudmik, 2013). Application-based mathematics instruction is focused on developing mathematics skills through the authentic lens of applying mathematics in real-world, problem-solving situations. Rather than teaching discrete, procedural skills in isolation within the Tier 1 context, we suggest that through the interweaving of the practices described in this chapter, educators create an application-based instructional framework (figure 3.2). Beyond this chapter and book, however, the critical work will be the experiences and reflections of each teacher in each classroom and the collective sharing of the learning through collaborative teacher experiences.

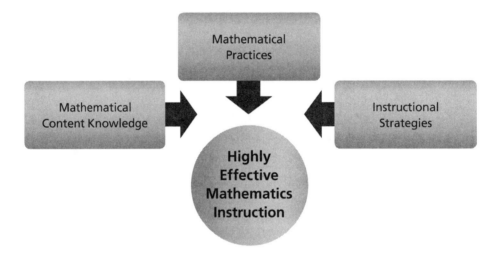

Figure 3.2: Elements of application-based mathematics learning.

Mathematical practices or processes are prioritized features of highly effective 21st century mathematics instruction. These practices reflect the critical *soft skills* of mathematics that have long-standing significance within mathematical learning and include communicating, modeling, persevering, and reasoning. While teachers may have implicitly worked to develop these skills, next-generation standards have launched them into the forefront of our efforts to improve mathematics instruction. The eight habits of mind identified in the CCSS, which we mentioned briefly in chapter 2, should be seamlessly woven into learning experiences from kindergarten through grade 12.

The practices provide common language and goals for all teachers of mathematics. In addition, the development of these goals from primary grades through high school allows teachers to deepen students' learning and guide the application of mathematics concepts and procedures. These practices are neither new (see Kilpatrick et al., 2001; NCTM, 2000) nor found solely within the CCSS, but are an integral part of many provincial and state mathematics standards.

The CCSS (NGA & CCSSO, 2010d) mathematical practices were developed by combining the NCTM (2000) process standards of problem solving, reasoning and proof, communication, representation, and connections, and the strands of mathematical proficiency specified in the Mathematics Learning Study Committee's report *Adding It Up* (Kilpatrick et al., 2001). Those strands are:

> adaptive reasoning, strategic competence, conceptual understanding (comprehension of mathematical concepts, operations and relations), procedural fluency (skill in carrying out procedures flexibly, accurately, efficiently and appropriately), and productive disposition (an inclination to see mathematics as sensible, useful, and worthwhile, coupled with a belief in diligence and one's own efficacy). (NGA & CCSSO, 2010c, p. 6)

These combined proficiencies represent one of the most significant features of next-generation standards and have the potential to have the greatest impact on transforming mathematics education. It is simply impossible to teach in a traditional manner and integrate these instructional practices. Changing standards without changing practices will have little impact on student achievement.

The integration of these eight practices and related reasoning-based habits of mind holds significant promise for increasing student engagement, confidence, and achievement in mathematics. While students must master concepts and skills and apply them in problem-solving situations, it is the fluid nature of these practices and processes that will empower students for a lifetime of creativity, critical thinking, and communication related to mathematical problem solving, establishing them as 21st century mathematicians.

These eight mathematical practices require rich, varied instructional approaches. The exclusive use of teacher-directed instruction will not allow students to fully practice and develop mastery of these practices, nor will the exclusive use of student-directed learning experiences be sufficient. Teachers will need to strategically integrate a variety of experiences to allow all students the opportunity to employ these practices.

The following is a brief overview and explanation of these eight practices.

Make Sense of Problems and Persevere in Solving Them

The NMAP (2008) research indicates that student beliefs about learning impact mathematics achievement to a greater extent than natural ability. The report specifically recommends that teachers and educational leaders consistently reinforce the positive impact effort has on mathematics achievement. This mathematical practice combines this principle with the expectation that students make sense of real-world problems, find entry points or first steps in solving the problem, and persevere as they test various approaches. This may represent a shift from traditional approaches in which teachers provided the formula or computational strategy to solve the problem. However, teachers must continue to monitor students for frustration and intervene strategically. For example, if a teacher notices a student is disengaged from a problem and overwhelmed, asking the following questions can scaffold the student to success without compromising his or her learning experience: Can you remember any other problems that were like this one? What did we do first? What happens when you try that? Eventually, the student will internalize these reflective questions to promote his or her metacognitive processing.

In addition, students develop plans for solving problems and flexibly refining approaches as necessary. It is critical that teachers provide students at the earliest grades with opportunities to wrestle with problems and discover various solution paths that successfully solve problems. When developing this mathematical practice, teachers should ask well-developed questions to prompt thinking, as opposed to simply providing answers or models. This practice not only fosters flexible critical thinking, but also students' willingness to persevere and try various solutions to a problem. The willingness to work through difficult problems is a skill that will serve students across all content areas and throughout their lives. The following statements describe this practice in student-friendly language for grades 6–8. When working on a challenging problem:

- I will restate the problem in my own words and draft a plan
- I will stick with it and not give up
- I will monitor my work to make sure it makes sense and change my plan if it doesn't

Reason Abstractly and Quantitatively

Students should be provided with multiple opportunities to apply concepts and skills within a variety of contexts. Quantities must be represented abstractly, and situations must be represented symbolically; numbers and mathematical expressions should be translated to verbal situations, and verbal situations should be translated to numbers and mathematical expressions. Students should understand the meaning of quantities and the relationship of the quantities to the situation and be able to flexibly use the properties of operations. While this is a complex task, elements of the practice should be embedded in instruction even in the earliest grades. The following statements describe this practice in student-friendly language for grades 6–8.

- I can use numbers, words, or manipulatives to show I understand a problem.
- I can create a story about a number problem to make sure it makes sense to me.
- I can use words to contextualize a number problem.
- I can use numbers to see if my solution is reasonable.

Figure 3.3 provides a sample of a student application of this practice.

Essential learning outcome: Solve real-world proportional and percent problems

Mathematical practice in student-friendly language:

- I can use numbers, words, or manipulatives to help me show I understand a problem.
- I can use numbers to help me check to see if my solution is reasonable.

$$\frac{\$1.00 \text{ US}}{\$1.0867 \text{ Canadian}} = \frac{\$x \text{ US}}{\$2{,}000 \text{ Canadian}}$$

"If I had \$1.00 of U.S. money, I would get about \$1.09 in Canadian money. That means if I had \$1.00 Canadian money, I would get less than \$1.00 U.S. That helps me know my answer to this problem will have to be less than \$2,000 to be reasonable."

Figure 3.3: Reasoning abstractly and quantitatively with proportions.

Construct Viable Arguments and Critique the Reasoning of Others

Students should develop the ability to justify solutions, communicate their ideas and strategies to others, and respond and revise their work based on questions and critiques from others. Even our youngest learners should begin their mathematics journey with multiple daily opportunities to talk, draw, and write about mathematics. They must have the opportunity to explain their reasoning and justify their thinking using appropriate language. Students who are proficient with this practice are able to compare the viability of plausible arguments and determine if their conclusions are logical or flawed. If their conclusions are flawed, students should be able to ask questions of themselves and others to clarify and refine those conclusions. In classroom cultures that promote a growth mindset and model this type of metacognition and dialogue, students develop a willingness to take mathematical risks and discover innovative strategies.

Across all grade levels, students can develop the ability to listen to, or read, the explanations of others, decide if the explanations make sense, and recognize elements of successful solutions. When students have the opportunity to listen to the reasoning of others and examine the reasonableness of their thinking, they can examine the validity of solutions and gain from others' perspectives. They can validate their own thinking against alternate solutions, thereby deepening their knowledge of the mathematics. These experiences develop skills invaluable for mathematical learning, as well as learning in other content areas. The following statements describe this practice in student-friendly language for grades 6–8.

- I can develop logical mathematics solutions and explain them to others.

- I can listen to others' mathematics solutions and determine if they are reasonable.

Model With Mathematics

Students should become proficient in applying the mathematics they know to solve problems in everyday life. Mathematics learning should not be done in isolation from real-world applications. If instructional activities are intentionally designed to connect to everyday life, students will begin to see mathematics in their world. For instance, as Myers Middle School seventh-grade students learn to solve real-world proportional and percent problems, they begin to attach meaning to this concept by comparing prices of popular items in proportional terms.

Students are able to model relationships between problem and solution through the use of diagrams, two-way tables, graphs, and equations. Examples of proportional real-world problems that would promote student modeling of the mathematics may include the following.

- Determine the amount of rice to cook based on serving size and number of guests.

- Determine if a sale that was 3 for $19.99 was a better value than $6.75 each with a 10 percent discount.

This practice, stated in student-friendly language, follows.

- I can use the mathematics I know to solve problems in everyday life.

- I can use mathematics to describe what's happening in a situation.

Figure 3.4 provides a sample of a student application of this practice.

Essential learning outcome: Solve real-world proportional problems

Mathematical practice in student-friendly language:

• I can use the mathematics I know to show how I solve problems in everyday life.

$1.00 US	=	$x US
$1.0867 Canadian		$2,000 Canadian

"To solve this problem, I have to determine the conversion rate from Canadian dollars to U.S. dollars. This would be approximately .92. To solve this problem, it isn't efficient to use manipulatives, so I am going to multiply .92 times $2,000. I know that an appropriate tool would be a calculator, but even without the calculator I can estimate that .90 of $1,000 is approximately $920. I have to double this because I had $2,000. This means the conversion should be around $1,840."

Figure 3.4: Mathematical modeling with proportions.

Use Appropriate Tools Strategically

Within each grade level, students should develop a thorough understanding of available mathematics tools, including technology-based resources and the availability, benefits, and limitations of the tool in relation to a specific situation. Simply stated, while educators should explicitly introduce and model the use of mathematics tools (such as calculators, virtual manipulatives, algebra tiles, fraction bars, graph paper, grid paper, hundreds charts, and spreadsheets), students should receive multiple experiences of exploring the use of varied tools in various contexts. It is important to note that many of these tools exist through technology-based apps. For some students, technology-based tools may have a positive impact on the learning experience. We should avoid predetermining the manipulatives needed in mathematical learning experiences and should instead consider what options may match the learning and be prepared to prompt students through questioning. Students who determine their own choice of tool demonstrate an understanding of the mathematical utility and purposes of various tools. Students should be provided opportunities to make independent choices based on mathematical contexts and personal learning needs. They should also be able to determine when a tool presents a limitation and when a combination of tools may best meet their needs. The following statement describes this practice in student-friendly language for grades 6–8.

• I can identify the most effective mathematical tools and technology I need to solve a problem.

Attend to Precision

Students are able to recognize the need for precision in different situations; they are able to identify when precise answers are required versus when approximate answers are appropriate. For instance, if the ratio at which a sports drink mix is added to water is 1.5 / 5, it is most critical to identify the units—1.5 ounces versus 1.5 cups mixed with 5 cups or 5 gallons creates great differences in the quality of the drink! An exact or precise identification of unit is necessary for this real-world problem. There are also mathematical situations in which an estimate represents the need for minimum allowable quantity. For example, if providing two water bottles for each member of a sports team, having extras (more than the minimum) is acceptable whereas not having enough (fewer than the minimum) would be problematic.

Students also must develop the ability to communicate precisely to others; this includes stating the meanings of mathematical symbols. Miscommunications ensue if there is confusion about purchasing 500 grams of cheese versus 500 kilograms of cheese. Proficiency in this practice includes the ability to carefully identify the units of measurement, label axes, and calculate accurately and efficiently. Precision also involves the consistent and appropriate use of symbols (for example, congruent versus equal signs, $x = 4$ instead of just putting 4 as an answer). The following statements describe this practice in student-friendly language for grades 6–8.

- I can solve problems with precise answers or know when an estimated answer is appropriate.

- I can clearly communicate about mathematics with mathematics vocabulary.

Look for and Make Sense of Structure

Students should develop the ability to identify patterns in mathematical problems and recognize how numbers and shapes are organized and combined. To make sense of structures that support efficient problem solving, students will need to mentally sort through numerous concepts and skills and determine which structures help organize or simplify a problem. For instance, when students are able to recognize how various attributes (shape, angles, number of sides) determine how shapes are organized and how those attributes determine relationships, they can utilize those skills to simplify and solve problems. To determine the surface area of a cylinder, students can identify the figures that comprise the surface area of a cylinder (see figure 3.5).

This recognition will allow them to process the structures to solve the problem instead of simply relying on the use of a procedure or abstract formula ($A = 2\pi r2 +$

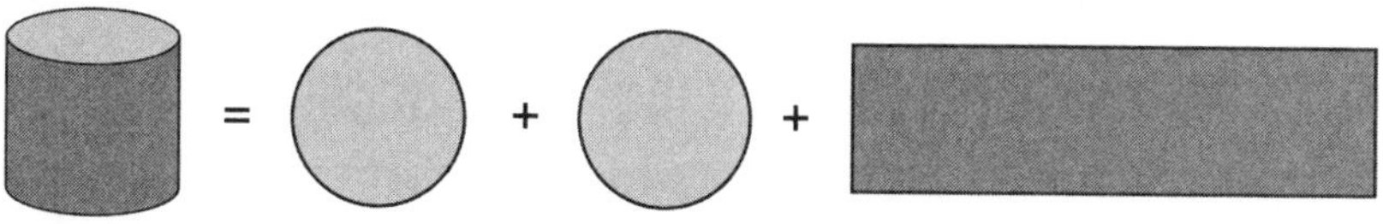

Figure 3.5: Shapes that comprise the surface area of a cylinder.

$2\pi rh$). While students should also be able to recall, create, or recognize the formula, the process of recognizing the structures and relationships between the shapes represents a more authentic understanding of the problem. Figure 3.6 shows an example of an activity in which students are finding and using structures when thinking proportionally.

Essential learning outcome: Solve real-world proportional problems.

Mathematical practice in student-friendly language:

- I can use the structures of mathematics to solve problems in different ways.

As the collaborative team at Myers Middle School develops the conceptual unit for proportional reasoning, they recognize students must develop a fluid understanding of multiplicative relationships in order to be able to utilize the structural elements of proportions. They present the following task to students.

"A concert marketing company is planning merchandise products and sales for an upcoming music festival. Data from previous festivals are used to order merchandise to sell. T-shirts with the festival logo have been a top seller. Purple fitted tees have been the top seller for females, and the traditional black T-shirt has been the top seller for males. An analysis of last year's ticket sales revealed that for every twenty-two males, there were twenty-five females. Approximately 75% of those attending the concert bought shirts. If 2,000 tickets are sold, approximately how many of each shirt would you recommend the marketing company stock for the festival?"

A fluid understanding of the multiplicative relationships would provide students the ability to recognize and employ the following sample structures to solve the problem:

Part/Part

$$\frac{22 \text{ males}}{25 \text{ females}} \quad \text{or} \quad \frac{25 \text{ females}}{22 \text{ males}}$$

Whole/Part

$$\frac{47 \text{ attendees}}{25 \text{ females}} \quad \text{or} \quad \frac{22 \text{ males}}{47 \text{ attendees}}$$

Part/Whole

$$\frac{25 \text{ females}}{47 \text{ attendees}} \quad \text{or} \quad \frac{22 \text{ males}}{47 \text{ attendees}}$$

Figure 3.6: Finding and using structures when thinking proportionally.

The following statement describes this practice in student-friendly language for grades 6–8.

- I can use the structures of mathematics to solve problems in different ways.

Look for and Express Regularity in Repeated Reasoning

The ability to recognize that operations or calculations are repeating or following a predictable pattern is important mathematically. The ability to recognize repeated reasoning is critical as students begin to draw connections between mathematical skills and concepts. When dividing 100 by 3 or converting 1/3 to a decimal, students should recognize the repeating decimal as they make sense of the solution. In addition, as students explore real-life mathematical problems, like converting from Celsius to Fahrenheit, repeated reasoning should help them construct a formula or efficient strategy for solving the problem without reliance on a predesigned formula. As students develop proficiency with this practice, they will generalize strategies to a variety of mathematical contexts and construct effective and efficient shortcuts to solve increasingly complex tasks. In fact, it is through the use of each of these practices that students become genuinely proficient with thinking, problem solving, and communicating mathematically. Figure 3.7 shows an example of an activity in which students use repeated reasoning when thinking proportionally. Note that this is the same problem that is posed to the students in figure 3.6. This illustrates how teachers can leverage the same concept or problem to represent multiple practices.

The following statement describes this practice in student-friendly language for grades 6–8.

- When calculations repeat or follow a pattern, I can use those patterns to create formulas and algorithms to solve the problems more efficiently.

The adoption of Standards for Mathematical Practice or of closely aligned next-generation state and provincial standards promises to significantly increase mathematical achievement in the United States and Canada. However, significant attention must first be dedicated to richly embedding mathematical practices in all mathematical learning. Mathematical practices serve as powerful catalysts for deep learning. These practices, however, must be directly linked to the mastery of prioritized standards. By integrating practices and reasoning with exploration of content, teachers and students will move beyond the procedural elements of mathematics to considerations of the thinking and applications involved in developing a true mastery of mathematics. The following are professional development activities to facilitate the integration of mathematical practices.

Essential learning outcome: Solve real-world proportional problems.

Mathematical practice in student-friendly language:

- When calculations repeat or follow a pattern, I can use those patterns to create formulas and algorithms to solve the problems more efficiently.

As the collaborative team at Myers Middle School develops the conceptual unit for proportional reasoning, they recognize students must develop a fluid understanding of multiplicative relationships in order to be able to utilize the structural elements of proportions. The repeating patterns associated with proportions can be recognized through the use of ratio tables. They present the following task to students.

"A concert marketing company is planning merchandise products and sales for an upcoming music festival. Data from previous festivals are used to order merchandise to sell. T-shirts with the festival logo have been a top seller. Purple fitted tees have been the top seller for females, and the traditional black T-shirt has been the top seller for males. An analysis of last year's ticket sales revealed that for every twenty-two males, there were twenty-five females. Approximately 75% of those attending the concert bought shirts. If 2,000 tickets are sold, approximately how many of each shirt would you recommend the marketing company stock for the festival?"

Males	22	44	88	176	352	704	1,408
Females	25	50	100	200	400	800	1,600
Total	47	94	188	376	752	1,504	3,008

Figure 3.7: Repeated reasoning when thinking proportionally.

- **Remove the title of each mathematical practice:** Teams of educators review the practice and develop a title and summary to share with the group.

- **Present the title of each mathematical practice by itself and match the description of the practice with the title:** Identify one way that each practice will impact instruction.

- **As grade-level teams review the mathematical practices, develop a summary of each practice and how it is relevant to grade-level expectations:** Prepare an explanation of what that mathematical practice would look like with students at your grade level.

- **Present grade-level summaries in a vertical team format:** Each vertical team discusses how the complexity of the mathematical practice increases across grade levels.

- **Review the mathematical practices:** Identify practices that align with current standards. During an upcoming time frame, plan for targeted implementation of two to four practices that align with prioritized standards. Teams bring examples of student work or anecdotal observations. How can integration of these practices be enhanced in the future? What did professionals find particularly effective?

- **Review standards or outcomes for a targeted grade level:** Choose a mathematical practice to embed in instruction that addresses the standard.

As education communities begin the transition to the use of next-generation standards, teams should devote careful consideration to mathematics standards and practices in order to plan for high-quality implementation. Integration of mathematical practices, whether provincial mathematical processes or the NCTM's (2000) mathematical principles, will help shift mathematical learning from being solely procedure based to including all three elements of mathematical mastery: conceptual understanding, procedural understanding, and application.

Promoting 21st Century (Application-Based) Mathematics Learning

In addition to embedding the identified mathematical practices, professional learning communities and classroom teachers must also explore, identify, and utilize highly effective (evidence-based) instructional strategies for mathematics. The impact of next-generation standards on mathematical achievement will be minimal unless district leadership teams, learning communities, and classroom teachers reflect on and improve instructional strategies for mathematics. Adopting new standards without embedding mathematical practices and utilizing evidence-based instructional strategies will not impact student achievement to the extent that we hope. To realize our goal of increased student mastery of mathematics, revisions to our practice must include adjustments to how we teach. We must facilitate student learning of critical content through integrating mathematical practices with learning experiences that utilize highly effective instructional practices and strategies. These evidence-based instructional strategies must promote deeper understanding and fluid application of mathematics. Figure 3.8 illustrates the interrelated factors that promote highly effective teaching and learning. If any of these three elements are missing from unit and lesson design, the effectiveness of the mathematics instruction will be compromised. However, if teachers proactively integrate these three elements, students will experience a rich 21st century mathematics learning environment.

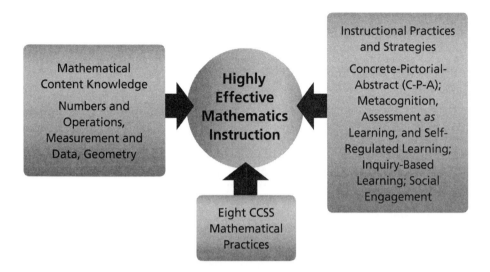

Figure 3.8: Interrelated factors that promote highly effective teaching and learning.

Simply stated, teachers are the most critical link to student learning. Their understanding of mathematical content and use of evidence-based instructional practices have significant impacts on student learning. As educators work to improve mathematics achievement, we have experienced the proverbial pendulum swings in mathematics instruction. We have been challenged to shift from teacher-centered to student-centered classrooms, from a computational focus to a conceptual focus. We have learned we need to differentiate, to include inquiry, to write about mathematics, to talk about mathematics, and to play with mathematics. We have been told we have too many standards and that they lack coherence, alignment, and rigor. We have been charged with creating opportunities for students to construct their own mathematical meaning and to apply mathematics in real-world situations.

We also recognize the critical importance of the role of the teacher in the classroom (Slavin & Lake, 2007). Instructional design and decisions guided by the professional expertise of educators are critical factors that can and do overcome many of the obstacles to learning that students face (Marzano, 2003). Students enter our schools with a diverse understanding of early numeracy. Some have experiences that include counting games and rich conversations about bigger, taller, more, and less. Others have had very limited language and mathematical opportunities. The complexity of student needs is unquestionable. In order to promote learning for all students, educators need a variety of practical, effective instructional practices that include differentiation strategies based on student learning needs. There is a great deal of research regarding evidence-based instructional practices and strategies for mathematics. Table 3.2 provides a list of instructional practices that should be a part of all mathematical learning experiences. The challenge for educators is to manage the organization of practices and targeted strategies that connect to specific concepts and skills.

Table 3.2: Instructional Practices for Mathematics

Inquiry-based learning: The opportunity for students to build on prior knowledge and construct new meaning. Students are provided the opportunity to wrestle with the mathematics.
Vocabulary instruction: Mathematics is a second language. Mathematics vocabulary needs to be explicitly taught in order for students to demonstrate precision and communicate about mathematics.
Think-alouds and think-alongs: Orally and visually modeling the cognitive processes that are occurring. Provides a model of what is happening in a mathematician's head during problem solving. Think-alongs represent a group think-aloud opportunity.
Metacognition: Learners monitoring their own learning and engagement. Self-regulated learning and assessment *as* learning are critical components.
Social engagement: Cooperative learning experiences including the sharing and defending of mathematical thinking. Includes evaluating the mathematical thinking of others.
Mathematics journaling: Students write and draw about mathematical learning. Can include metacognitive reflections.
Explicit instruction: Clear instruction provided through gradual release models: I do (teacher model); we do (students and teacher together); you do it together (students work with one another to build, practice, and solidify new learning); you do it alone (students apply learning with corrective feedback).
Fluency building: Provide students enough practice to build automaticity with basic facts and the fluent application of skills that can be applied to new contexts.
Real-world learning and project-based learning: Opportunity for students to apply mathematical learning in a real-world context.
Formative assessment: Ongoing process in which teacher and student learning is monitored in order to adjust instruction.
Concrete-Pictorial-Abstract (C-P-A): Student learning related to mathematics concepts is introduced and practiced with concrete objects, whenever possible. Learning is transferred to a pictorial representation that may include drawings, virtual manipulatives, tally marks, or graphs. Learning is finally transferred to abstract application that includes algorithms and mathematics symbols.

No single instructional practice or program will meet the needs of all learners. Teachers need a variety of evidenced-based instructional practices to engage students in meaningful mathematics learning. While increasing teachers' use of instructional strategies through the use of graphic organizers, questioning techniques, journaling, and mathematics games is important, it is not sufficient. There is no scarcity of such activities within mathematics textbooks, materials, and websites. Adapting and organizing those activities into cohesive learning experiences, however, often represents a frustration for teachers. By utilizing research related to effective mathematics instruction, we can design mathematical learning experiences that are fluid and dynamic enough to promote rich student learning and integrate effective instructional strategies. Later in the chapter, we will propose four phases of mathematical learning that will provide practical structures for mathematics learning. First, however, we offer four instructional practices that should be utilized in every phase to provide differentiated, core instruction. The following evidence-based instructional practices should be interwoven throughout all phases of mathematics learning.

1. Concrete-Pictorial-Abstract (C-P-A) representations of content and learning

2. Metacognition, assessment *as* learning, and self-regulated learning

3. Inquiry-based learning

4. Social engagement

C-P-A Representations

C-P-A representations of mathematics learning is likely the most researched and validated instructional practice (Maccini & Ruhl, 2000; Witzel, Mercer, & Miller, 2003) for mathematics, and yet it continues to be underutilized or misused in many classrooms. Whenever possible, mathematical learning should begin within the context of concrete, authentic experiences. As our grades 6–8 students continue to explore and learn mathematics (to compute with rational numbers, solve multistep inequalities, and compute the surface area of a prism), the utilization of concrete, 3-D, hands-on interactions will promote a solid conceptual understanding of mathematics. For instance, the introduction of proportional reasoning may include the mixing of a sport drink or use of authentic coins and bills to demonstrate a currency exchange. Concrete explorations of concepts scaffold students to a deep conceptual understanding that enhances application and problem-solving abilities.

Beginning with concrete, authentic objects is critical to supporting the development of conceptual understanding. Students must then be scaffolded in their learning to create visual, 2-D representations of mathematics. Pictorial representations of mathematics may include drawings, tally marks, pictures, or other graphic displays. For some students, this may include written words telling the story of their learning. Students may also use digital technology including Prezi or StoryMaker to animate the mathematics. When students translate concrete learning into visual representations, they increase their understanding and ownership of mathematics. This process deepens their learning and serves as scaffolds to later, abstract representations of concepts. As students develop conceptual understandings of mathematical concepts and translate their learning from concrete, experiential learning to pictorial representations, abstract presentations of the concepts have more meaning than if the learning process begins with the introduction of a computational procedure, formula, or algorithm. Student utilization of equations or formulas to represent their learning and understanding of mathematics typically represents a more efficient strategy. However, the use of strategies without understanding does not promote fluid application of mathematics. C-P-A promotes the conceptual understandings that should be a foundational complement to equations, formulas, and procedures. Table 3.3 provides a sample list of concrete manipulatives and pictorial tools for some of the prioritized standards identified in chapter 2 for grades 6–8 students.

While the research base for utilizing C-P-A is substantial and long-standing, a new element of this instructional process that is receiving increased attention is the use of virtual manipulatives. Mathematics professor Patricia Moyer and colleagues (Moyer, Boylard, & Spikell, 2002) define *virtual manipulative* as "an interactive, Web-based, visual representation of a dynamic object that provides opportunities for constructing mathematical knowledge" (p. 372). Studies indicate that virtual manipulatives are as engaging and effective as concrete manipulatives (Olkun, 2003) and that the use of either concrete or virtual manipulatives increases students' understanding of mathematics over pencil-and-paper tasks alone (Brown, 2007). In their classroom-based research project, Kelly Reimer, classroom teacher, and Patricia Moyer (2005) demonstrate an increase in the understanding of fractions over pencil-and-paper instruction for students who use virtual and concrete manipulatives. They report a potential added benefit that the virtual manipulatives can provide a more-direct connection between concrete and abstract mathematical symbols. A similar research project (2011) integrated preservice teachers' conceptual understanding of mathematics with concrete and virtual manipulatives to increase conceptual understandings (Hunt, Nipper, & Nash, 2011). Candidates who used concrete and virtual manipulatives in alternating order reported that first using concrete manipulatives created greater conceptual understanding than beginning with virtual manipulatives. In instances where concrete manipulatives were used first, followed by virtual manipulatives, transitions between concrete understanding and abstract algorithms were improved. We suggest that concrete manipulatives be used to introduce and establish students' conceptual understandings and that virtual manipulatives be utilized to facilitate the transition between concrete conceptual understanding and the use of abstract algorithms.

For students motivated by technology, virtual manipulatives may increase engagement, motivation, and mathematics achievement. Virtual manipulatives may be more readily available at home than the concrete manipulatives in the classroom since virtual manipulatives are available through multiple online resources. As teachers monitor students' learning in response to integrating virtual manipulatives into

Table 3.3: Concrete Manipulatives and Pictorial Tools for Grades 6–8 Prioritized Standards

Mathematics Concept	Suggested Manipulatives	Pictorial Tools
Rational Numbers	Tiles, fraction bars, fraction circles, attribute blocks, Unifix Cubes, fraction squares	Number lines, number line bars, pictures
Integers	Algebraic tiles, integer chips	Number lines, thermometers, pictures or drawings
Surface Areas	Nets and tiles	Drawings and pictures
Fractions	Tiles, fraction bars, fraction circles, attribute blocks, Unifix Cubes, fraction squares	Number lines, number line bars, pictures
Decimals	Cubes, rods, flats, coins, Unifix Cubes	Place value charts and discs

the evidence-based C-P-A instructional process, it is likely that some students will greatly benefit from the integration of this technology. Students should be empowered to explore available tools, evaluate their usefulness in problem-solving situations, and determine which tools are most effective for them as learners.

Metacognition, Assessment *as* Learning, and Self-Regulated Learning

The terms *metacognition, assessment* as *learning*, and *self-regulated learning* are receiving a great deal of attention in education forums. Metacognition is commonly associated with an awareness of one's own thinking and learning process and is closely associated with both the assessment *as* learning process and self-regulated learning. Assessment *as* learning occurs as students monitor their progress, incorporating feedback to make adjustments as they develop metacognitive skills (Western and Northern Canadian Protocol [WNCP], 2008). In *Visible Learning for Teachers: Maximizing Impact on Learning*, Hattie (2012) reports that students' self-reporting of achievement had the greatest influence on student learning of any variable studied. This ranking was based on analyses of over twenty years of global research and validates the positive influence of metacognition on student learning. Self-regulated learning refers to a student's ability to direct thoughts and actions in order to achieve goals and respond to external and internal factors (Zimmerman, 2008). This requires learners to continually monitor and adjust their physical and mental statuses, including the ability to understand a task, resist distractors, and persist when tasks are difficult. Self-regulated learning involves motivation, metacognition, and strategic decision making (Winne & Perry, 2000).

While the terms *assessment* as *learning*, *self-regulated learning*, and *metacognition* are not synonyms, they are deeply intertwined and interdependent. In practical applications within classrooms, these terms describe very similar instructional and learning practices. Each focuses on learners' (our students' and our own) monitoring, reflecting, adjusting, and evaluating of personal learning experiences. For our purposes in addressing key instructional strategies for mathematics, we will focus on key metacognitive processes. These processes are critical to students' engagement with mathematics; students cannot be proficient with the mathematical practices or mathematics as a whole without utilizing metacognitive processes. As students develop mathematical understandings, they must also develop the ability to reflect on their own progress and identify the tools and strategies that are most supportive of their individual learning. Students should be given frequent (even daily) opportunities to process and think about their mathematical thinking. In other words, metacognition should be modeled for students, and students should be provided with chances to be metacognitive. A simple journal, like the one shown in figure 3.9, can provide students with this chance.

This means I can recognize the repeated pattern and use an equation to solve a problem.

Mathematics I already knew that helped me learn this is:

1. My basic mathematics skills
2. Understanding of rational numbers
3. How to compute with fractions

I can solve a real-life problem with proportions.

This is important because there are many opportunities to solve proportional problems (recipes, money exchange, sale prices, and so on).

To solve a proportion problem I represent the proportions as ratios or fractions.

= 1 representative

Equation: $6/1 \times 18/x$ (six people equals 1 representative, 18 people = x representatives)

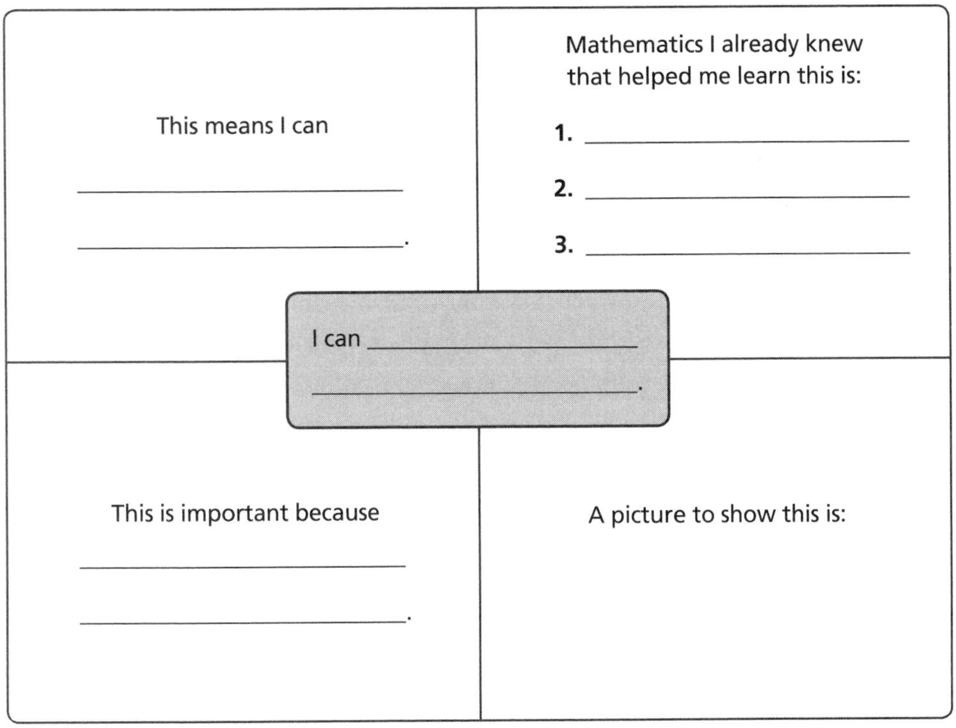

This means I can

_____

_____.

Mathematics I already knew
that helped me learn this is:

1. _____

2. _____

3. _____

I can _____

_____.

This is important because

_____

_____.

A picture to show this is:

Figure 3.9: Sample student journal—metacognition for mathematical learning.

While metacognition has traditionally been linked with *learn to learn* and *thinking about thinking* practices, like the use of concept maps, mnemonics, and study strategies, we propose that metacognition is much broader in its application to mathematical learning and should be modeled and explicitly connected to all elements of learning. The Singapore syllabus for mathematics defines metacognition within a mathematics context as (Yoong, 2002):

- Consistent and intentional monitoring of thinking while carrying out a task

- Seeking alternate strategies or approaches

- Monitoring the reasonableness of an answer

We believe this summary of metacognition encapsulates the intent of all mathematical practices and should be explicitly integrated into mathematics education from the earliest of ages. The modeling and practicing of these metacognitive processes would help create a classroom culture that accepts that mathematics can be messy and that multiple solutions and strategies are acceptable and valuable. In middle school classrooms, metacognitive practices can be modeled through teacher think-alouds and student think-alongs (whole-class or small-group think-alouds followed by sharing strategies, approaches, and ideas) (Yoong, 2002). Rather than begin instruction with a preplanned, presolved problem to model a solution step by step, instruction with mathematics concepts can begin with explorations of new concepts with connections to previous ideas. This should include the modeling and practice of the metacognitive skills that demonstrate that first approaches do not always work.

Researcher and author Alan Schoenfeld (1987) suggests that students can be taught a set of favorite questions to integrate into problem-solving activities to promote the enduring use of metacognitive mathematics practices. Questions include: What are you doing? Why are you doing it this way? and How does this help you? While teachers can adapt these questions, they should be introduced within the context of authentic problem-solving experiences (figure 3.10). We suggest teachers model these questions first within the context of a think-aloud and then within group-shared think-alongs. Students should then be guided to use these same or alternate favorite questions with each other and to use these questions to monitor their own and each other's thinking.

Our school is purchasing new sporting equipment for the soccer teams. We have $3,875 to spend. We have quotes from two different companies and have to decide which company is offering us the best value.

World of Soccer has offered the school a 15 percent discount on the total $3,875 order and free shipping.

Soccer Warehouse has offered the school a 15 percent discount on the first $2,000 spent and a 20 percent discount on everything over $2,000. Soccer Warehouse will charge a 3 percent shipping fee.

Adapted Favorite Questions	
Teacher Think-Aloud	**Student Think-Along**
• What am I going to do? • Are there tools that may help me? • Why am I doing it this way? • Do I still think this is the best way, or should I try something different?	• What are we going to do? • Are there tools that may help us? • Why are we doing it this way? • Do we need to change anything?

Figure 3.10: Favorite questions problem-solving model.

In addition to using metacognitive processes to monitor understanding and reasonableness in problem-solving situations, metacognition should be present in all phases of mathematical learning. For instance, when students are listening to peers or the teacher explain a possible solution or idea, students should check their own understanding and ask clarification questions as needed. As students demonstrate understanding and mastery of mathematics concepts, metacognitive practices can be used to self-assess learning. Metacognitive strategies can also be included in mathematics journaling. Mathematics journals (see figure 3.11) provide students the opportunity to reflect on critical mathematical terms, processes, previous mathematics learning that added to their understanding, and strategies that worked well for them as a learner.

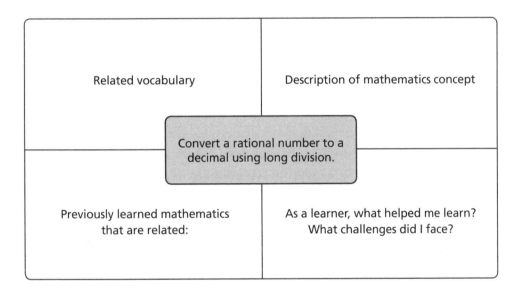

Figure 3.11: Mathematics journal used to reflect on specific concepts.

As students develop mathematical understandings, they must also develop the ability to reflect on their own progress and identify the tools and strategies that are most supportive of their individual learning. Metacognition, assessment *as* learning, and self-regulated learning are evidence-based instructional practices whose utility spans the grade levels and are critical in all mathematical learning contexts. While metacognitive skills are certainly applicable to all content areas, they are critical and have been underrepresented within mathematics and must be intentionally and explicitly taught. Emotionally safe classroom cultures are critical for high levels of learning and will be cultivated and nurtured through metacognition, assessment *as* learning, and self-regulated learning practices.

Inquiry-Based Learning

While metacognitive models provide examples of questions and prompt learners to continually monitor their understanding and engagement, teacher questioning is another evidence-based instructional practice that significantly impacts student engagement and understanding of mathematics. Effective questioning can be used to promote student thinking, extend student learning, and help students create connections. However, effective questioning is a complex skill, dependent on teacher understanding of mathematical concepts. If we have not given ourselves sufficient time to consider the concepts of prioritized content through the eyes of students, we may miss rich questioning opportunities. Too often, we invest significant time planning models, solutions, and explanations without also investing time developing effective questions that can promote greater learning.

Author Steven Reinhart (2000) coined the following phrase in his journey to improve questioning techniques as a mathematics teacher: "Never say anything a kid can say" (p. 478). He recognized that, as teachers who are committed to moving away from traditional mathematics instruction, we need to stop *telling* students how to solve problems. Instead, we should equip them with the skills and tools they need with carefully considered questions that prompt their thinking. While questioning techniques may seem second nature to educators, the reality is that effective questioning is dependent on conceptual understanding of mathematics and requires careful listening to students.

Quality questioning is deceptively simple and requires practice to optimize. Observational research comparing questioning techniques of U.S. teachers and teachers from China indicates that U.S. teachers are significantly more likely to ask closed questions (questions that have one right or wrong answer) and to go on to another student when an answer is incorrect. Teachers from China were observed to ask more open-ended questions such as: "Why did you try that? What could you try differently?" If student answers were seemingly incorrect, questions were used to clarify and prompt (Perry, VanderStoep, & Yu, 1993). As we develop and extend our questioning techniques, it is often helpful to predraft questions that would clarify student thinking or extend their learning. Questions to ask during inquiry or exploration learning experiences are:

- What do you recognize about this problem?
- Is this similar to any other problems we've worked?
- What mathematics skills do you already have that may help?
- What tools may help you? Why do you think that will be an effective or useful tool?

- How can you represent this problem: with a picture, manipulatives, a virtual drawing, or some other way?

Questions to help students who are experiencing difficulty are:

- In your own words, what is the problem?
- What parts of the problem can you draw or represent?
- What is the problem asking for?
- What information do you have?
- What have you done so far to resolve the problem?
- What mathematics strategy or tool may help?

Questions to check student understanding include:

- How does this relate to _____?
- Why is that true?

Questions to promote deeper understanding and problem solving are:

- Can you think of a time this would not work?
- How would you prove that?
- Can you create a mathematics rule for that and convince your team?
- Why is this important?
- What other strategy could you use?

Questions to summarize and reflect are:

- What new vocabulary did we use today?
- What will this mathematical concept or skill help us do?
- What are three key points from today's mathematics time?

Social Engagement

Encouraging social interaction for mathematical learning should span our students' educational experiences from kindergarten through graduation. Social learning to support mathematical learning is highly effective. Research documents that social and intellectual support from peers and teachers is associated with higher mathematics performance (NMAP, 2008). Students need extensive opportunities to discuss mathematics, to problem solve in collaborative teams, and to process alternative strategies other students use. Students also benefit from the encouragement of peers and the opportunities to take safe risks (taking a chance to share your ideas even when you don't know the exact answer) when problem solving. Additionally, peer and social interactions are a critical strategy for embedding the mathematical practices the Common Core State Standards recommend, NCTM process standards, and the National Research Council's (Kilpatrick et al., 2001) strands of mathematical proficiency.

We must also construct learning experiences that honor student differences. For instance, it is important that we give all students think time to process mathematics as an individual as well as with a peer or peers. Some students may do very well in groups of four while others need the safety of just one other peer. This differentiation can be critical to student learning. Another strategy to encourage the social interactions for students who tend to be more introverted or have expressive language difficulties is to provide the opportunity to first journal about their thinking or to develop a podcast. While those strategies can serve to scaffold toward greater levels of social dynamics, it should be our goal for all students to engage in actual face-to-face interactions to experience real-time feedback and develop questions that can deepen thinking and foster the drawing of connections. Students who converse about mathematics have an opportunity to reflect on their learning and to authentically use mathematical language. Exchanging ideas within a mathematics context, such as when two students share insights on the implications of a data set representing population shifts, provides a catalyst for questioning, discussions of differing views, and refinements of judgments. Even for our youngest learners, discussing various ways to sort or count objects promotes critical thinking and communicating about mathematics. As students listen to the mathematical reasoning of others, they must shift from their own individualized perceptions and understandings to consider the alternate insights of others.

Four Phases of Application-Based Mathematics Learning

Teaching and learning mathematics is multifaceted, and student needs are complex. We have explored three factors that contribute to highly effective mathematics instruction: teachers' conceptual understanding of the mathematics, integration of mathematical practices, and instructional practices. They represent three areas that influence our schools and classrooms on a daily basis. They also represent three areas that can significantly and positively impact student learning. As we explore using more of these practices in our classrooms, it is also important to give consideration to how to utilize these practices within a structure that promotes student learning. We recognize that there is no one way to present information so that all students master critical learning. We also recognize a need for structures that guide reflective and responsive teaching and learning. Mathematics instruction cannot be limited to a series of activities or games or the traditional teacher model / student practice approach. The NMAP (2008) indicates there is no evidence to support either student-centered or teacher-directed practices to the exclusion of the other.

Research-based instructional practices are employed within cognitively planned, pedagogically sound phases of mathematics instruction and learning. The following are the four phases of application-based mathematics learning.

Phase 1: Explore and Inquire—Students are challenged with an authentic problem situation that requires application of the targeted prioritized learning outcomes. During this learning phase, direct instruction is not provided. Rather students, individually or within small groups, draw on knowledge and problem-solving strategies to construct meaning within the context of the problem. Educators facilitate conversations, prompt with questions, and validate student thinking. Students share insights, ideas, and strategies. This phase may integrate multiple content areas and interconnected mathematics. At consistent intervals, groups debrief their progress, illustrating various strategies to solve the problem. Classroom culture should promote safe sharing of ideas. The following three phases provide multiple opportunities for student learning and provide differentiated, responsive learning experiences. Phase 1 begins to establish conceptual understandings as described in chapter 2 *and* connects mathematics learning to an application basis.

Phase 2: Teach and Explain—Based on observations, checks for understanding, and other formative assessment data, direct, explicit instruction is provided based on student needs. This includes correction of misconceptions, strategy instruction, and vocabulary instruction. Students share insights, ideas, and strategies. Note that this phase does not necessarily include all learners in a classroom. Students that demonstrate thorough mastery during phase 1 may be provided enrichment experiences or may benefit from parts of the instruction. Within the context of this phase, instruction is provided in direct response to student learning needs and integrates effective instructional practices listed in table 3.2 (page 68).

Phase 3: Practice—When teachers have evidence of readiness, individual students or groups receive adequate opportunities to practice the prioritized learning outcomes to become fluent. Students share insights, ideas, and strategies. This phase targets the interconnection of conceptual understanding and procedural competency as described in chapters 1 and 2.

Phase 4: Demonstrate Ownership—Students demonstrate *ownership* of the prioritized learning outcomes. This may occur through the formative assessment processes in phases 1 or 3; this is not necessarily a separate task or test. Teachers may also design and administer assessment tasks that provide students an opportunity to demonstrate mastery. In every phase, it is important for students to share insights, ideas, and strategies in small groups, with partners, or through digital media. This phase will likely connect and link directly to the learning experiences from phase 1 and should integrate multiple interrelated mathematics concepts. It must be noted that it is predictable that some of our learners will not master the learning outcomes, despite the high-quality learning experiences the phases provide. What we learn about our students, including data we collect, should inform the design of interventions. Intervention instruction and assessment are discussed in greater detail in chapter 5.

While these four phases provide an overarching structure, they are not meant to be rigid four-day lessons; nor are they meant to serve as the structure for a sequentially completed, one-day lesson. Based on the learning context, some phases may be combined or unnecessary. Some students within a class may need to progress through all four phases while other students may need only aspects of the phases. In addition, the learning context will define the time each phase of instruction takes. For example, the explore-and-inquire phase may be a twenty-minute activity or it may be extended across two class sessions. These decisions require the professional insights of educators who understand the unique learning needs of each group of students. Teachers constantly check for student understanding, provide immediate and specific corrective feedback, and release students to greater levels of independence—all in relation to mastery of the prioritized learning outcomes.

This is RTI in action—the extent to which students are *responding* to *instruction*. The four phases of application-based mathematics instruction and learning serve as a flexible framework to guide student learning through the seamless weaving of mathematical content, mathematical practices, and instructional practices and strategies. Within this suggested framework, educators continually reflect on student learning to adjust instruction in response to student needs. These phases represent a fluid, dynamic, and responsive instructional model for mathematics. Each phase allows for differentiation, student-led discourse, and authentic integration of technology. We also recommend that instructional time, ideally every class session, includes dedicated time for students to share their thinking. It is not necessary that every student or every group share every time; however, teachers should strategically identify sharing opportunities that will extend thinking, celebrate learning, and address misconceptions.

In the following sections, we take a more in-depth look at each of the four phases.

Phase 1: Explore and Inquire

Imagine a classroom where students are learning a new mathematics concept or skill. Where is the teacher? Who is doing the talking? What technology is being employed? Perhaps the teacher is sitting with a small group of students, there are four other students sitting at a table with a variety of mathematics counters, two more students appear to be drawing on a tablet, and three other students are gathered around a laptop computer making notes on a sketchpad. Remember, this is the introductory phase of learning a new mathematics concept. The teacher has posed a question or problem that integrates what students have previously learned but not everything they need for this problem.

What we've come to recognize through our observations, classroom experiences, and research is that perhaps the best way for students to begin a new learning journey—to begin to master new mathematical content—is for the new mathematical content *not* to be explicitly taught. There is a place for explicit direct instruction, but there are significant benefits to students and teachers when students struggle with new knowledge.

- They learn perseverance and stick-with-it-ness.

- They begin to attach new knowledge to existing schema.

- Teachers gain authentic, timely views into students' existing knowledge and misconceptions.

- Teachers have immediate, real-time understanding of students' level of knowledge related to prerequisite skills and any misconceptions that may hamper understanding of the new prioritized mathematics standards.

Developmental psychologist Lev Vygotsky (1978) demonstrated that student learning is positively impacted by social interactions that allow students to connect new content to what they already know and that provide challenges and cognitive dissonance to facilitate new connections. Connecting what students already know to what students are ready to learn allows them the opportunity to construct their own meaning. Research suggests that student engagement and motivation are fostered when teachers present tasks that build on the students' prior knowledge and guide them to the next level. This process allows students to think through concepts for themselves (Hiebert & Stigler, 2004; National Center for Education Statistics [NCES], 2003).

Phase 1 of mathematical learning is designed to be a time in which students use prior mathematical knowledge to engage in new mathematical learning. Little direct teaching is done at this point in the instructional process; it is a time for students to grapple with mathematics ideas. Instructional activities should be based on real-world contexts and applications of prioritized outcomes and skills. Students should be asked to determine what tools may assist them and should have access to technology, manipulatives, artifacts, and prior learning. It is critical that teachers carefully monitor student discussions and frequently check for emerging student understanding, approaches that students use, and misconceptions that exist—in other words, to formatively assess student progress. What vocabulary are students using? What errors are impacting success? Which students require additional support? Which students are ready for more complex challenges or to take more responsibility for their learning? Information learned about the class, small groups, and individuals informs the following phases of learning. It may be determined that there is no need for whole-class explicit instruction related to this prioritized learning outcome, or it may be that the whole class requires direct instructional time for a mini-lesson to solidify prior skills or build background knowledge.

Travis's Grade 7 Class

Earlier in this chapter we saw how Travis Port and the seventh-grade teachers at Myers Middle School identified using proportional reasoning to solve real-world problems as an essential learning outcome for the first term. Most of Mr. Port's students have demonstrated mastery of multiplication, can compare quantities, and understand part-whole relationships. Mr. Port recognizes that he has three students who are not yet able to demonstrate an understanding of multiplication as repeated addition and two students who struggle with comparing quantities. As he plans for the introduction of using proportional reasoning, he combines his knowledge of proportional reasoning (content knowledge), mathematical practices, instructional practices, *and* his students' learning needs to design the explore-and-inquire phase of learning.

A casual observer may not recognize the complex differentiated structures that teachers design to promote student learning. Let's consider Mr. Port's grade 7 classroom. During the explore-and-inquire phase, the class attempts to solve the following:

> A national animation company is developing plans to build an amusement park in your area. One of the attractions simulates being shrunk to 1/10 your size. In order to create that visual, they will be building some props proportionally larger. Choose two classroom items and explain precisely, including projected measurements, how you would build them to scale to simulate being shrunk to 1/10 their size. Identify how you can visually present your findings.

While Mr. Port generally groups students with different skills together for phase 1, he is grouping students with similar learning needs for this critical skill. For example, the students who need concrete models work together, and students who have already demonstrated mastery of this skill work together to apply their learning to a real-world application. In addition, Mr. Port knows which of his students fluidly use metacognitive processes to make sense of the problem, and he knows which students continue to need scaffolding to organize their understanding and develop a plan. After a brief whole-group introduction of the task, eighteen of Mr. Port's students begin work on the task, while the other nine join him at a small-group area for think-along time to more thoroughly process the task and develop an initial plan. They ask each other questions such as "What approaches do you think you may try?" Mr. Port's understanding of content, tasks, and his students has empowered him to proactively plan differentiated supports that create a more challenging dynamic for some students and scaffolded support to build confidence for others as they engage in the explore-and-inquire task. Differentiation within the context of phases 1 through 4 is critical for student success. While there is not a focus on direct instruction in relation to the prioritized learning outcomes, there will be authentic opportunities for teachers to promote the mathematical practices, extend students' thinking, and identify mathematical strategies that should be modeled and shared. Mr. Port, for instance, recognizes that during the explore-and-inquire activities, his class is building proficiency with several of the mathematical practices. As table 3.4 indicates, multiple mathematical practices are applied during the learning experience, exemplifying that the mathematical practices are not separate from learning outcomes but rather integrated into the learning experience.

Table 3.4: Application of Mathematical Practices

Mathematical Practice	Class Application of Practice
Make sense of problems and persevere in solving them.	Students individually use metacognitive processes to check for understanding and identify connections to previous learning.
Reason abstractly and quantitatively.	Students recognize that their task is to order the numbers and to compare the value of the numbers.
Construct viable arguments and critique the reasoning of others.	Students have individual think time and group discussions to develop a plan and to try solutions. Within the context of a group, they justify their thinking and evaluate ideas of others to determine the group plan.
Model with mathematics.	Students share ideas and strategies within their small group and to the whole group. Myers Elementary strategically identifies different students on a rotating basis to share their thinking so they can check for understanding.
Use appropriate tools strategically.	Students must identify and choose what mathematics tools to use.

continued →

Attend to precision.	Students use the vocabulary of ones, tens, and hundreds to clearly communicate the value of numbers.
Look for and make sense of structure.	Students use what they know about place value and using concrete manipulatives or pictorial representations as strategies to compare and order the numbers.
Look for and express regularity in repeated reasoning.	Students note that place values always contain the digits zero to nine and that each place value to the left has a value ten times the place value to the right.

Phase 2: Teach and Explain

We believe that structured, supported student exploration of mathematics concepts and skills is a critical element of the mathematics instruction necessary to promote deep understanding and fluid application of mathematics. We also know, as validated by NMAP (2008), that many students need high-quality explicit instruction that is guided and informed by formative assessment practices to develop conceptual understanding and procedural competencies. Phase 2 of our instructional framework is the strategic teacher-led instructional time based on student needs. Teachers' careful observation of student thinking and application of concepts during phase 1 is critical for the planning of phase 2. Teachers may identify a mathematics strategy that naturally connects to a concept or skill that students should acquire to facilitate future learning. For instance, as students are exploring adding fractions with unlike denominators, they may demonstrate proficiency at using fraction bars, as illustrated in figure 3.12. In the first row, the fraction $\frac{1}{3}$ is represented in two different ways; in the second row, $\frac{1}{2}$ is represented in two ways.

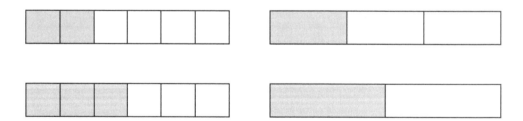

Figure 3.12: Fraction bars representing equal fractions.

While using fraction bars demonstrates the need to create like denominators and the identification of using a mathematical tool to solve a problem, a teacher may note that a student is not able to flexibly identify other strategies to solve the problem. Given the importance of fractions, teachers may determine that demonstration of this critical skill requires that the student have multiple means to solve problems that require determining like denominators. Instructional decisions within the context of the four phases of mathematics instruction could include providing direct instruction to increase students' repertoire of mathematics strategies.

Phase 2 is also an element of mathematics instruction that addresses the need to define, model, and learn critical vocabulary. Given the prevalence of content-specific terms, many consider mathematics a second language. Certainly with the increased focus on student modeling and communicating about mathematics, it is important to dedicate adequate instructional time to ensuring that students acquire the necessary vocabulary. Subtle changes in the vocabulary used between classrooms and support teachers can impede learning. In the instructional process, teachers should model consistent, student-friendly language and use graphic organizers to promote learning, including providing examples and non-examples.

Within the context of Myers Middle School's conceptual unit related to using proportional reasoning to solve real-world problems, the teacher team has identified key words and agrees on general definitions. These terms include *ratio, rate, proportions, part-to-part ratios, part-to-whole ratios,* and *units.* As Mr. Port observes students in phase 1, he identifies that all students require that these terms be clarified. It is not uncommon for students who have solid grasps of mathematical concepts to require direct instruction related to key vocabulary terms. Dismissing the importance of this work can interfere with students' abilities to communicate their understanding of mathematics or contribute to future misconceptions. Phase 2 should include the modeling of the mathematical practices through think-alouds and think-alongs. Students should be actively engaged in this phase of learning. Learning and insights should be shared through journaling and Share the Work time. Simple prompts may include "Today I connected our learning to . . ." and "Today, my aha moment was when . . ."

Phase 3: Practice

Proficient mathematics students must gain fluidity with mathematical concepts and procedural competencies. This includes the ability to process a problem or mathematical situation and sorting through mathematical knowledge to identify the strategies, algorithms, and tools needed to solve the problem. This level of mathematics processing requires that students practice skills and concepts enough to fluently apply them to various situations. One of the major criticisms of mathematics instruction in North America is that we cover too much curriculum too thinly; we do not provide our students with enough time and practice to develop fluency and ownership of mathematical learning.

The amount of time dedicated to phase 3 is dependent on student learning needs within a classroom. When designing practice opportunities, teachers should use authentic learning experiences to the greatest extent possible. There are also motivating and engaging games that provide students the opportunity to build fluency without the drill-and-kill worksheet experience. As we consider evidence-based practices for our learners, we must plan experiences that help develop fluency when applying concepts and skills within problem-solving situations and with basic facts and calculation strategies. A student's ability to quickly and efficiently retrieve basic addition, subtraction, multiplication, and division facts has a direct impact on his or her ability to complete more complex mathematics problems. When students are not fluent with these facts and rely on nonefficient means to compute, the cognitive processing of more complex mathematics is interrupted. As part of England's national curriculum, pupils receive a daily numeracy lesson with a focus on mental and oral calculation. Since the introduction of this practice, England has made the greatest advance in mathematics achievement by fourth-grade pupils of any country TIMSS sampled (Mullis, Martin, Foy, & Arora, 2012). In North America, there continues to be debate among educators regarding memorization of facts. Some schools and curricula dismiss the importance of developing fact fluency; however, intermediate and secondary teachers repeatedly identify fact fluency as a critical skill (Kirschner, Sweller, & Clark, 2006; NMAP, 2008). Conceptual understanding of number and facts is equally critical. We do not believe that the debate should be *if* students should develop fluency with facts, but *how*.

Phase 4: Demonstrate Ownership

As students develop the concepts, procedures, and applications related to mathematics learning, rich assessment practices should be employed that provide students an opportunity to demonstrate their learning in a variety of contexts. The focus of this phase is to provide students an opportunity to demonstrate their mastery of their learning. Demonstrations of learning include both informal and formal assessments, allow us to evaluate student learning and celebrate success, and identify ongoing needs for whole classes, small groups, and individual students. Assessment cannot be the sole responsibility of teachers. Students should self-assess their learning by asking themselves, "What are my strengths, goals, and next steps?" They should be prepared and empowered to communicate their learning to peers and teachers. Teachers should also reflect on student learning: "What are my students' demonstrated strengths, suggested goals, and suggested next steps?"

Demonstration of learning must not be limited to an end-of-unit summative test. Through student think-alouds, mathematics journaling, and students modeling or demonstrating mastery of mathematics, teachers can identify those who are demonstrating proficiency. Teachers' careful observation of students performing these tasks provides critical formative assessment information that can guide the development of instruction to meet various student needs. Phase 4 may occur for some students during any of the other phases. Application-based projects, for example, allow students to demonstrate mastery of a single concept or skill and can combine multiple mathematical concepts and other content areas.

Within the context of the four phases of mathematical learning, the NMAP (2008) recommends that students' experiences include a balance of student-centered and teacher-directed instruction. The four phases also present educators with a framework for organizing their best efforts for student learning and for integrating the eight CCSS (NGA & CCSSO, 2010d) mathematical practices or other state or provincial mathematical processes and evidence-based practices that include metacognition, social engagement, and constructivist experiences, for example. As opposed to haphazardly combining a series of activities, teachers can proactively organize effective, reflective, and responsive learning experiences that will promote student learning in Tier 1. Table 3.5 provides an example of the four phases of learning outlined by Myers Middle School to address the prioritized learning outcome. This table can also be adjusted to use as a lesson organization tool. Table 3.6 (page 90) provides an overview of how the framework can be used for planning by instructional teams.

Table 3.5: Sample Four-Phase Lesson

Instructional Goal: Use Proportional Relationships to Interpret and Create Scale Drawings			
Phase 1: Explore and Inquire	**Overview of Activity or Activities**	**Addressing Student Needs**	**Potential Teacher Questions**
Share the Work time—Students share a summary of their learning; choose a representative number of students to share or have all students share with other students.	Students choose two items from the classroom, for example, a desk and a chair. Students are to identify the size to scale and plan how to represent with modeling clay or other materials.	Grouping decisions—Create small groups (two to four students who have similar instructional needs). For example: Alayna and Claire—Extend their learning by having them create two different-sized representations. Rachel, Mackenzie, and Ashley—Provide direct teacher support at the beginning of the activity. They will also need step-by-step guidance. Devin—Provide an iPad to assist with communication needs and as a means to demonstrate understanding.	What tools do you think will help your team? What mathematics do you know that will help? What can you already tell each other about this problem? What mathematics vocabulary is important to share our strategies?

continued →

Phase 2: Teach and Explain	Overview of Activity or Activities	Addressing Student Needs	Potential Teacher Questions
Vocabulary to teach includes ratio, rate, proportions, and units. Share the Work time—Students share a summary of their learning; may choose a representative number of students to share or have all students share with other students.	Definitions are developed collaboratively with the teacher team for ratio, rate, proportions, part-to-part ratio, part-to-whole ratios, and units. Provide mini-vocabulary lessons for all students together. Provide an explicit guiding lesson modeling drawing to scale.	All students receive vocabulary instruction. Students choose how to record information to demonstrate their understanding. Preteach definitions to Rachel, Mackenzie, and John. To provide an extra challenge for Alayna and Ashley, they should predraft what they think the vocabulary words and definitions will be. Model and provide think-aloud examples to Rachel, Mackenzie, John, Steven, Randle, Devin, Ashley, and Cassady. All other students move to Phase 3: Practice.	How would you explain each vocabulary term to a friend? What picture can you draw to help you remember the meaning of the word? Where would you start to solve the problem? What would you do next? What tools would you use?

Phase 3: Practice	Overview of Activity or Activities	Addressing Student Needs	Potential Teacher Questions
Share the Work time—Students share a summary of their learning; choose a representative number of students to share or have all students share with other students.	Randle Golf Academy is designing a teaching tool that allows new golfers to understand the distance and angles to each hole. They have asked our class to create a map drawn to scale and a 3-D model. Everyone in your group will need to be able to explain the scale to an instructor from the Academy. Each member of your team is responsible for at least two holes.	Grouping decisions are made based upon student needs. Claire, Liam, and Alayna work collaboratively together, and their models will include sand traps and water hazards. Mackenzie, Devin, and Rachel design the first three holes and are given a step-by-step checklist. John and Steven need to work as partners, as larger teams overwhelm both of them.	What tools do you think will help your team? What mathematics do you know that will help? What can you already tell each other about the project? What mathematics vocabulary is important to this work?

Phase 4: Demonstrate Ownership	Overview of Activity or Activities	Addressing Student Needs	Potential Teacher Questions
Share the Work time—Students share a summary of their planning and learning; choose a representative number of students to share or have all students share with other students.	Exciting news! Randle Golf Academy was so impressed with our work, they would like to give each of you an opportunity to design a hole at their new teaching facility. You can include one sand trap or a water hazard. The instructors can't come to our classroom this time (it's golf season), so you will need to explain your hole through PowerPoint, Prezi, or with a drawing and written description. Make sure to include as many of our vocabulary words as possible in your project. Also, please use the math you learned during this unit to identify at least one other job we could do for Randle Golf Academy. Use proportions, part-to-part ratio, and part-to-whole ratios in this proposal.	Each student should complete this task independently as an assessment (demonstrate ownership). However, the task should be adapted for each student. John will utilize the assistive technology, including communication apps on his iPad. All students will be provided a rubric to encourage reflective self-assessment.	What math or technology tools were most helpful to you in this activity? What math strategies will help you? Describe your plan to me. Check your thinking with a partner. What connections can you make to this problem from other learning we've done?

Table 3.6 Instructional Plan

Instructional Goal: Clearly Identify the Prioritized Learning Outcome			
Phase 1: Explore and Inquire	**Overview of Activity or Activities**	**Addressing Student Needs**	**Potential Teacher Questions**
Authentic, challenging problem situations—Engaging students in the learning creates excitement, motivation, and interest in the mathematics.	Describe what the activity is and what materials are needed.	Identify any grouping decisions. What students should work together? Should students be grouped with similar needs or diverse needs? What students may need preteaching? What students may need practice with a prerequisite skill? Other differentiation needs?	What questions may prompt student learning? Help connect learning? Deepen student learning?
Phase 2: Teach and Explain	**Overview of Activity or Activities**	**Addressing Student Needs**	**Potential Teacher Questions**
Explicit instruction, gradual release, think-aloud, modeling, scaffolding, vocabulary instruction—Level of instruction should be based upon student needs identified in phase 1.	Describe what the activity is and what materials are needed.	Identify any grouping decisions. What students should work together? Should students be grouped with si milar needs or diverse needs? What students may need preteaching? What students may need practice with a prerequisite skill? Other differentiation needs? Decisions should reflect formative assessment information from phase 1.	What questions may prompt student learning? Help connect learning? Deepen student learning?
Phase 3: Practice	**Overview of Activity or Activities**	**Addressing Student Needs**	**Potential Teacher Questions**
Practice should be adequate enough to build automaticity.	Describe what the activity is and what materials are needed.	Identify any grouping decisions. What students should work together? Should students be grouped with similar needs or diverse needs? What students may need preteaching? What students may need practice with a prerequisite skill? Other differentiation needs? Decisions should reflect formative assessment information from phases 1 and 2.	What questions may prompt student learning? Help connect learning? Deepen student learning?

Phase 4: Demonstrate Ownership	Overview of Activity or Activities	Addressing Student Needs	Potential Teacher Questions
Ownership and assessment, project-based learning, authentic assessment—students *justify* their solutions.	Describe the options students have to demonstrate mastery of skills.	Identify any grouping decisions. What students should work together? Should students be grouped with similar needs or diverse needs? What students may need preteaching? What students may need practice with a prerequisite skill? Other differentiation needs? How will each student demonstrate understanding if working in a small group? Decisions should reflect formative assessment information from phases 1, 2, and 3, scaffolded support, and preteaching needs.	What questions may prompt student learning? Help connect learning? Deepen student learning? What questions may identify misconceptions?

Conclusion

Within the RTI framework, Tier 1 is the foundational proactive learning experience for students. Our ongoing efforts as educators are to identify the most effective instructional approaches to promote student learning for the purposes of continuously improving instruction and assessment. The process of partnering with students in the journey of mathematical learning is a complex task for elementary educators, who often do not feel confident in their own mathematical abilities. We recognize that teachers' understanding of the mathematics that they are working to ensure student mastery of impacts student achievement. Teachers' understanding of mathematics topics (concepts, procedures, and applications) empowers their ability to design tasks that engage students and connect to prior learning. Their understanding of the mathematics also allows them to develop questions that challenge and clarify student thinking and identify potential misconceptions.

In addition to teachers' understanding of the mathematics, integration of mathematical practices, or habits of mind, promotes deep authentic learning of the mathematics. As students engage in those practices, including modeling with mathematics and constructing detailed, viable arguments, they cognitively engage in deeper learning of the mathematics. When students shift from their own problem-solving strategy and evaluate the logic of others' solutions, their understanding of mathematics and the ability to apply knowledge deepen.

Teachers' use of evidence-based instructional practices represents the professional skill and knowledge that provide the structure and opportunity for student learning. Within a positive learning environment that promotes a culture of problem solving and a willingness to try strategies that may not work, students become effective problem solvers. Within this context, students learn to think metacognitively, assess their strategies, and monitor their own learning. Teachers must utilize explicit supports for those students who require direct instruction to master prioritized skills. Teachers can only responsively plan and design instructional activities to meet the diverse needs of learners through effective formative assessment strategies.

Through the integration of teachers' thorough understanding of mathematics and mathematical reasoning and effective use of instructional practices and strategies, students' engagement, confidence, and mastery of mathematics will flourish. The instructional practices and flexible, responsive instructional phases presented in this chapter are supported by a strong research base that develops students' conceptual understanding, procedural competency, and ability to apply mathematics through problem-solving opportunities. These application-based learning experiences promote deeper levels of learning and require that adequate time be dedicated to address the prioritized standards.

The next chapter provides resources for teachers to implement in Tier 2 and Tier 3 intervention. Despite our best efforts at core Tier 1 instruction, some students will require supplemental supports. Proactively planning and preparing for tiered interventions in mathematics is a necessity if we hope to achieve our goal of all students mastering the skills required for college and a skilled career.

Chapter 4

Tiers 2 and 3 Approaches to Foundational Conceptual Understandings and Mathematical Practices

Most middle school mathematics teachers understand mathematics deeply and thoroughly. In our conversations with grades 6–8 teachers, however, they are candid about the challenges of meeting the instructional needs of students who are missing foundational skills. Teachers find it quite challenging to address these early skills within the context of their classrooms or even to identify how to model those concepts and skills simply because they are so far removed from the grade-level content.

In this chapter, we first describe how middle school teacher teams may organize and deliver Tier 2 supports for students. We provide examples of common formative assessment results that would identify the need for Tier 2 supports and examples of how a few prioritized middle school concepts may be taught differently. We then examine how middle school leadership and RTI teams may organize and deliver Tier 3 supports for students. We will provide examples of universal screening evidence that may reveal significant deficits in foundational mathematics skills as well as examples of the types of intervention strategies that would explicitly target diagnosed needs.

A Review of RTI

As educators plan for implementation of RTI, they must continuously and strategically improve Tier 1. The implementation of RTI for mathematics requires a commitment to reflect and refine Tier 1 instruction, especially with the support of collaborative teams. Tier 1 will always require that we differentiate and scaffold teaching so that students also receive the levels of support they need to maximally grow within core instruction. We also recognize some students will require more time and alternative strategies within a unit or after the completion of a unit to master prioritized content. This is Tier 2. Moreover, we fully appreciate that some students will enter classrooms needing intensive Tier 3 interventions in order to access grade-level content. Within the context of our grades 6–8 classrooms, this will include students who have significant mathematical deficits. These deficits may represent primary mathematics skills that are far below the expectations of middle-level classrooms.

Before we further discuss Tier 2, it's important that readers have a clear understanding of the definitions of Tiers 2 and 3. Tier 2 and Tier 3 address different types of student need. Tier 3 does not simply represent a more rigorous version of Tier 2 (Bloom, 1968, 1984). Tier 2 represents proactively planned periods within the school day or school year during which more time and alternative approaches (or enrichment with tasks of greater depth and complexity) can be provided to students for whom short-cycle assessments reveal the targeted need. Tier 3 represents intensive, targeted interventions necessary because significant deficits in foundational prerequisite skills exist. We believe that many schools and school districts have been frustrated in their efforts to successfully implement RTI-based systems of support because of a misunderstanding about the very definitions of Tiers 2 and 3.

The Importance of Tier 2

Most schools have designed systems of support that address Tiers 1 and 3—we must proactively embed Tier 2 as well. Often, the struggles with effective RTI implementation stem from skipping over Tier 2. If high-quality instruction isn't working, there is a rush to go directly to Tier 3 and look for outside-the-classroom assistance. The consequences of not providing Tier 2 include the following.

- Students may catch up on foundational skills because of the quality Tier 3 support, but in the absence of Tier 2, they risk failing to keep up with the current year's priorities.

- Students are not ready for the next unit, course, or grade level.

- Student motivation is likely to wane.

- We do not deliver on our commitment to a growth mindset—we do not structure buffers that allow more time to master the priorities of Tier 1 through alternative strategies and approaches.

The aim of Tier 2 is to provide more time and alternative approaches with the goal of mastering prioritized Tier 1 content. We might have Tier 2 (buffer) time between units of instruction (akin to Benjamin Bloom's models; see Bloom, 1968, 1984) or within each school day. In both scenarios, the approach is the same—intervention or enrichment related to the previous unit's priorities. Tier 1 classrooms should be heterogeneously grouped; Tier 2 groupings can be homogenous—students can be specifically grouped within the Tier 2 block based on their specific needs, the mastery, or lack of mastery of a focused set of unit priorities.

Let's imagine that a seventh-grade end-of-unit common assessment during the first semester reveals that a few students have not mastered the addition and subtraction of integers. A few students, for example, produced the solution featured in figure 4.1.

Evaluate the expression −15 + 23 − 8 + 9	
Show your work.	Explain each step.
−15 + 23 − 8 + 9	I rewrote the problem.
15 + 23 − 8 + 9	I changed the sign of the first number, following the rule.
38 − 8 + 9	I added 15 and 23.
30 + 9	I subtracted 38 and 8.
39	I added 30 and 9.

Figure 4.1: Metacognition in mathematics.

When the seventh-grade teacher team met to analyze student learning, they noticed students' errors and planned the Tier 2 intervention and enrichment that they would provide during RTI time—a dedicated block of time during which students receive Tier 2 intervention or enrichment depending on evidence of need. This is used in addition to what is embedded during instruction. While core instruction began in the teachers' heterogeneous classrooms on a new unit, the team would organize students by need, as determined by common assessment evidence, during this intervention time. Students who did not master addition and subtraction with integers would benefit from more time and alternative strategies with the goal of learning these prioritized concepts. Students who demonstrated mastery would benefit from enrichment—collaborative experiences with tasks of greater depth and complexity related to integers.

The following sections describe a sample Tier 2 intervention activity for the topic of integers. Teachers first assess students' ability to perform skills including estimating and justifying, calculating, imaging, and applying. The following figures provide examples of outcomes that indicate to the teacher whether an intervention is needed. The number sense outcome question used in this example to assess these skills is $-24 + 8$. After reviewing students' answers, teachers then provide needed interventions for these skills in a short lesson. Examples of how those interventions might be implemented appear after each figure. Finally, the student is asked to reflect on the effectiveness of the intervention.

Estimate and Justify

The student should be able to use a grade-appropriate strategy to estimate and justify an answer (see figure 4.2).

Estimate ($-24 + 8 = $)	
Intervention Not Needed	$-20 + 10 = -10$ (The student rounds -24 to -20 and 8 to 10, and is able to explain his or her thinking.)
Intervention Needed	$-20 + 10 = 30$ (The student misunderstands negative integer relationship to positive integer.)

Figure 4.2: Teacher team's analyses to inform Tier 2 estimation needs.

The student's response will aid the teacher in helping the student move to a more sophisticated estimation strategy.

Using manipulatives or pencil and paper, ask:

- What does -20 look like?
- What could you have in real life that looks like -10?
- What could you have in real life that looks like 20?

If the student is still struggling, bring your own examples to the intervention such as temperature, money, or measurement to assist with the conversation.

Calculate

The student should use a grade-appropriate strategy to solve the problem. He or she may show multiple ways or nontraditional methods to come to an accurate answer (see figure 4.3).

Calculate (−24 + 8 =)	
Intervention Not Needed	• Using 24 − 8 to show difference between two numbers but also recognizing that the answer is below zero and giving the answer a negative sign • Using a number line to show −24 and counting along to −16 to calculate the answer
Intervention Needed	• Adding of the two numbers together to 32 • Having the numerical answer correct (the difference between the two numbers is 16) but not recognizing the answer will be below zero so not indicating a negative sign before the 16

Figure 4.3: Teacher team's analyses to inform Tier 2 calculation needs.

The student's response gives insight into his or her thinking and helps the teacher intervene with an appropriate lesson.

To assist this student with recognizing that the answer is below zero:

• Use a number line and counting, first counting down to −24 and then counting up by 8.

• Use empty egg cartons (2) to represent the −24. Add 8 eggs. How many eggs are we short to fill the carton?

Image

The student should be able to represent the mathematics using a sketch or diagram (see figure 4.4).

Image (−24 + 8 =)	
Intervention Not Needed	Using a number line to represent growth from below zero Showing a large circle (representing a hole) being filled in by a smaller circle
Intervention Needed	Drawing 24 dots and then 8 dots

Figure 4.4: Teacher team's analyses to inform Tier 2 visual modeling needs.

To assist the student with imaging strategies, we might have the student redraw the twenty-four dots as holes to represent the negative and then fill in eight of the holes.

Apply

The student should be able to explain how to use the mathematics in a real-life situation (see figure 4.5).

Apply (−24 + 8 =)	
Intervention Not Needed	"The baker was supposed to have made two dozen donuts but only eight cooked properly. How many is he short?"
Intervention Needed	"I use this in math class."

Figure 4.5: Teacher team's analyses to inform Tier 2 application needs.

To assist the student with application strategies, we would, in small groups, have students discuss where they would see the number 24 in real life (hockey player's season total of goals, liters of fuel for a trip, and so on). Continue the discussion and have students create real-life stories for −24 + 8. Have the students share their ideas with the class—the more creative the better!

Reflect

The student should be actively conscious of his or her learning (see figure 4.6).

Reflect (−24 + 8 =)	
Intervention Not Needed	Student is able to articulate what he or she found easy or hard, what he or she could do next time to improve with justification, or both.
Intervention Needed	Student is unable to justify response.

Figure 4.6: Teacher team's analyses to inform Tier 2 self-reflection needs.

To assist the student with reflection strategies, we would prompt the student to recall intervention effectiveness (Bird & Savage, 2014). For example, What was difficult? What was easy? What will you remember for next time?

At the conclusion of the new unit, members of the seventh-grade teacher team administer the end-of-unit common assessment that measures learning of the unit's priorities of Tier 1. In addition, students who had not mastered the preceding unit's priorities—adding and subtracting integers—are also assessed with alternate but parallel problems from the preceding unit's common assessment (see figure 4.7).

Evaluate the expression 8 + −12 − 10 − 2	
Show your work.	Explain each step.

Figure 4.7: Sample assessment task—evaluating expressions.

The re-administration of the portions of the previous unit's common assessment on which students had not yet demonstrated mastery measures the extent to which students are responding to Tier 2 intervention.

Tier 3

As previously noted, students need not fail to respond to Tier 1—and then Tier 2—before we provide Tier 3. Tier 3 is a different type of support altogether. Tier 3 interventions are intensive, targeted supports to ameliorate significant deficits in reading, numeracy, and behavior. It's entirely possible for students to receive differentiated instruction at Tier 1 as well as intensive Tier 3 intervention if significant deficits in foundational skills exist. This would assume that teachers are successfully differentiating and scaffolding at Tier 1 and that evidence from common formative assessments does not indicate the need for more time and alternative supports (also known as Tier 2) to master the prioritized standards at Tier 1. We caution against providing the opportunity for only Tier 2 *or* Tier 3 but not both.

The consequences and costs of providing Tier 3 interventions to students with significant deficits in foundational skills while other students have access to Tier 2 are very real—students may catch up by gaining proficiency with foundational skills, but they risk falling behind with this year's priorities. It's unlikely that a school's schedules allow for Tier 3 intervention in ELA and mathematics without students temporarily "missing" another subject. But this is the cost of ensuring success in the foundational skills of literacy, numeracy, and behavior when significant deficits exist.

In these cases, we believe it's educational malpractice not to insist upon providing Tier 3 interventions in reading, numeracy, and behavior in place of important social studies, science, and elective opportunities, a position supported by Allington (2009) and Fuchs and Fuchs (2007). Reading, basic mathematics, and self-regulating behavior are requisite for any success. The inability to do these often precludes success in other content areas and in life.

Education policy consultant Craig D. Jerald (2009) sums up the importance of ensuring students develop these basic skills thusly:

> The demand for educated workers will continue to be high, and those who obtain postsecondary education or training can continue to expect to earn a premium while those who do not will have far fewer opportunities to earn a living wage. (p. 30)

President and CEO of the Association of Private Sector Colleges and Universities, Steve Gunderson (2013), further suggests there will be 55 million new jobs in the United States by 2020 and that 65 percent of these jobs will require some level of postsecondary education and training. Clearly, a student who lacks in the foundational skills will all but eliminate the opportunity to complete high school graduation requirements and, concomitantly, access any training beyond school. The potential outcomes are clear. Without prioritized Tier 3 interventions, students with significant deficits will not finish high school, they will not fully participate in the comprehensive high school experience, or they will not graduate ready for college or a skilled career. They will not lead a productive life if they do not possess foundational literacy, numeracy, and behavior skills.

Before we give an example of the type of Tier 3 interventions that may be necessary for students with significant deficits in foundational skills, let's describe and define the concepts that compose the foundational skills for success in middle school mathematics.

A review of next-generation elementary mathematics curricular standards across North America reflects three domain areas in which to focus mathematical learning in kindergarten through grade 5 to ensure conceptual clarity (including key vocabulary terms) for teachers and their learners. These three broad domains are:

1. Numbers and operations

2. Measurement and data

3. Geometry

These domains represent the building blocks of learning that become increasingly more complex and interrelated in middle school, and represent foundations that are critical to accessing the essential mathematics content in grades 6–8. These concepts and skills form the same basis for mathematical learning as the five domains— phonological awareness, phonics, fluency, vocabulary, and comprehension—form for reading. Mathematical domains have their own progression of interconnected skills. As interventions are designed to correct deficits in these areas, we must be aware of how these various elements of mathematical learning are interconnected with grade- level learning. For instance, fractional awareness that is developed in K–5 connects to proficiency in proportional reasoning, algebra, geometry, personal finance, and measurement.

A thorough understanding of foundational mathematics domains and the pro- gression of critical, foundational concepts and skills that comprise each of them is required for educators to be empowered to:

- Develop engaging, challenging learning opportunities that promote rich mathematical learning

- Identify evidence-based instructional practices that appropriately connect the learning needs of students to the identified prioritized learning outcomes

- Utilize a variety of assessment strategies to gather the evidence that allows for the adjustment of instruction to meet each student's unique needs

- Respond in *real time*, as the learning is occurring in the moment, with questions that clarify potential misconceptions and deepen learning with immediate and specific corrective feedback

- Differentiate Tier 1 instruction for students with varied current levels of readiness

Table 4.1 (page 102) describes in greater detail the foundational skills developed in the numbers and operations domain and that should be considered in the design of interventions for grades 6–8. In addition to summarizing the primary foundational mathematics skills, the table identifies both concrete and pictorial tools that middle school teachers can use to address them.

Table 4.1: Numbers and Operations—Concepts and Tools for Intervention

Mathematics Concept	Suggested Manipulatives	Pictorial Tools
Number Sense	Cubes, blocks, buttons, Unifix Cubes, color tiles	Tens frames, dot cards
Place Value	Cubes, rods, flats, Unifix Cubes	Hundreds charts, place-value charts and discs
Addition and Subtraction	Unifix Cubes, blocks, buttons, crayons (any hands-on manipulative)	Hundreds charts, number lines
Multiplication and Division	Color chips, cubes, tiles, buttons	Times tables, area models
Fractions	Tiles, fraction bars, fraction circles, attribute blocks, Unifix Cubes, fraction squares	Number lines, number line bars
Decimals	Cubes, rods, flats, coins, Unifix Cubes	Place-value charts and discs

Learning Outcomes for Interventions

Just as we noted the need to identify, clarify, and define prioritized learning outcomes in chapter 2, we should utilize a similar process to identify, clarify, and define essential learning outcomes for interventions. We should collaboratively and collectively provide descriptions that define mastery of these standards and the critical mathematical terms. For instance, composing and decomposing numbers would likely be considered a priority foundational concept or an essential learning outcome that requires intervention when students don't have mastery. We must therefore define standards such as "Students compose and decompose numbers to 100 or beyond" in specific terms. Are students expected to use decomposition in operations with whole numbers, decimals, or fractions? What specific language or vocabulary terms will be utilized? This would undoubtedly be most productively defined within the local context and in terms of viable goals for intervention blocks. It's important to note that all stakeholders—including teachers and interventionists—should use consistent language and have consistent interpretations of conceptual understandings to avoid misconceptions and possible frustration for learners, especially the vulnerable learners who require intervention.

As collaborative teams identify which students require supplemental intervention through a convergence of assessment data (classroom assessments, common assessments, universal screeners, and diagnostic tools), they must clearly define what skills and concepts are worthy of intervention time, including needed depth of knowledge, potential misconceptions, and mathematical vocabulary. The best intervention is a targeted intervention at both Tiers 2 and 3.

Conceptual Understandings Related to Numbers and Operations

Numbers and operations is a domain that encompasses foundational mathematics skills, including the mastery of the use and understanding of quantities and the ability to manipulate quantities through operations to solve problems. A thorough mastery of numbers can be equated with phonological awareness, one of the critical elements of learning to read (Weber, 2013). Just as phonological awareness, the umbrella term that encompasses the recognizing, understanding, and manipulation of spoken sounds, is the basis for reading and written expression, the understanding of numbers has the same critical, foundational application for mathematics. This knowledge and understanding is often referenced as number sense. The NMAP (2008) describes number sense at the fundamental level as the ability to fluently identify the numerical value associated with small quantities and, at the more advanced levels, as knowledge of numbers written in fraction, decimal, and exponential forms. Number sense encompasses the precursor skills for all subsequent mathematical learning, including algebra, measurement and data, geometry, and statistics. Number sense also serves as the basis of the mathematical knowledge critical for independent adult life, including postsecondary education and skilled-career readiness.

The depth of number sense that students must acquire to become proficient mathematically goes beyond rote counting and number recognition. Mastery of numbers, which must receive significant focus in the primary grades, includes students establishing a meaning for numbers that is connected to real-life experiences and recognizing relationships among numbers (Charles & Lobato, 2001). The study of numbers as outlined in most next-generation standards includes the ability to fluidly apply counting strategies and recognize quantity and comparisons. A fluid understanding of number sense should integrate conceptual understanding with procedural fluency and problem solving (NMAP, 2008). Deficits in any of these factors in relation to any of the elements of number sense can contribute to future difficulties in mathematical learning, which relates directly to many of the learning challenges middle grade teachers observe. Figure 4.8 (page 104) identifies the elements of number sense. Developing these types of knowledge packages can assist teacher teams in identifying key terms and concepts that must be addressed through intervention to develop conceptual understanding necessary to close mathematical learning gaps. In addition, these tables remind us that our students must master three facets of mathematics: (1) conceptual understanding, (2) procedural understanding, and (3) application in order to be able to apply the mathematics to real-world, problem-solving situations. Simply stated, each is critical and vital for application-based mathematics. In addition, the tables, through the use of a vertical arrow, remind us that many standards have discrete objectives that reflect a progression from simple to complex.

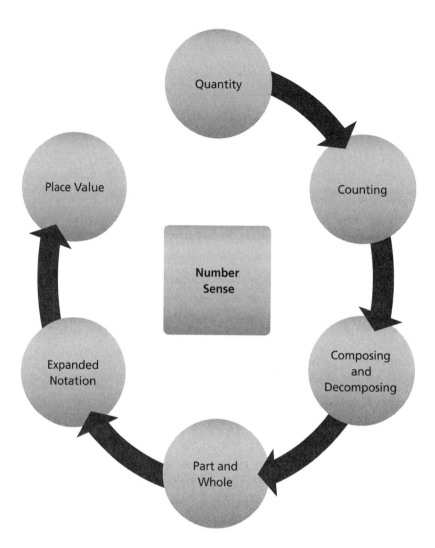

Figure 4.8: The elements of number sense.

Number sense includes many concepts that, while not generally part of the grades 6–8 curriculum, are significant hurdles for many middle school students. The knowledge package featured in figure 4.8 can assist teacher teams in identifying key terms and concepts that are necessary for developing conceptual understandings related to number sense. Table 4.2 provides sample vocabulary and concepts related to foundational numeracy skills with which middle school students may have significant deficits that necessitate Tier 3 interventions. This table may assist teacher teams in identifying what foundation skills need targeted and intensive intervention. The table is organized by grade-level bands to provide middle-level teachers a broad context of student deficits and to recognize that the nature of primary mathematics learning includes the need for concepts to develop over time. For each term or concept, student mastery must exist at the conceptual understanding, procedural understanding, and application levels.

Table 4.2: Key Concepts—Number Sense

Concept	Description
	Conceptual Understanding → Procedural Understanding → Application
Correspondence	One-to-one relationships between two groups, often demonstrated by touching individual concrete objects with or without using the counting sequence, or by matching two objects. Earliest application can include real-life activities like setting the table and putting on gloves.
Count	Ability to recite numbers or to identify the number of objects, often demonstrated by rote oral counting or identifying the number of objects in a group. Instruction includes cardinal numbers (one, two, three . . .) and ordinal numbers (first, second, third . . .).
Subitize	Ability to see and know the number of objects without counting. As students master the relationship between numbers, five and ten serve as benchmarks. Five-frames and ten-frames are appropriate instructional tools to strengthen this skill.
Conservation	Ability to retain in memory the number of objects without recounting and to recognize that the quantity of a group does not change even if appearance or order changes. A critical skill for using the counting-on strategy.
Cardinality	The principle that when counting, the last number used is the total number of objects.
Compare and Contrast (Magnitude)	Ability to identify if a group or number is greater than or less than another group or number. Learning should include comparison language and requires one-to-one correspondence.
Place Value and Numeration System	The number system that allows students to move from individual counting to more efficient groupings of tens, hundreds, thousands, ten thousands, and so on. The position that a number occupies identifies its value.
Compose and Decompose Whole Numbers	Within the place-value system, the ability to recognize that a series of numbers maintain their individual values and the resulting combined (or added together) value. In the earliest grades, this should focus on the power of ten. Beginning in kindergarten, groups of two and five become the basis of composing and decomposing. By grade 1, learners are composing and decomposing based on sets of ten. In grade 2, students work toward using expanded form to compose and decompose $(400 + 20 + 5 = 425)$. Numbers can be taken apart based on the values of the numerals $(425 = 400 + 20 + 5)$. Flexible application would include the use of words and sets (4 hundreds + 2 tens + 5 ones = 425).

The left side of the table is labeled, from bottom to top: **Complex ← Simple**

continued →

Complex ← Simple	**Addition and Subtraction**	Addition: An operation that joins groups to make larger groups (I have 5 baseball caps, and a friend gives me 2 more caps: $5 + 2 = 7$) or represents part + part = whole (I have seven baseball caps: 5 black and 2 white: $5 + 2 = 7$).
		Subtraction: An operation that removes a part of a group to create a smaller group (I have 7 baseball caps, and I give 2 to a friend: $7 - 2 = 5$).
		OR identifies a comparative amount (I have 7 baseball caps; you have 2 baseball caps. How many more caps do I have? $7 - 2 = 5$).
		OR identifies a missing addend (I had 5 baseball caps, and my friend gave me some baseball caps. Now I have 7 baseball caps. How many did my friend give me? $7 - 2 = 5$).
	Multiplication and Division of Whole Numbers	Multiplication: An operation where students learn to efficiently complete repeated addition patterns. May represent total of a set of equal groups (three groups of students with five students in each is fifteen students: $3 \times 5 = 15$). May represent total of arrays (six rows of eggs with two in each row: $6 \times 2 = 12$). May represent total number of combinations (number of ways three pairs of shoes and five pairs of socks can be combined: $3 \times 5 = 15$). May represent rate (Lori has three books; Cal has five times as many books as Lori: $3 \times 5 = 15$).
		Division: An operation where students learn to efficiently complete repeated subtraction of equal groups. May represent number of groups of a given size in a total (15 students total with 5 students in a group equals 3 groups: $15 \div 5 = 3$). May represent an equal share of a total number (fifteen apples shared into three equal groups is five apples for each group: $15 \div 3 = 5$). May represent the number of columns or rows in an array (twelve bottles of water in three rows is four columns: $12 \div 3 = 4$).
		Like addition and subtraction, multiplication and division are reciprocal operations. Conceptually sound, strategy-informed practice with basic multiplication and division creates automaticity, which supports fluid problem solving.
	Fractions and Unit Fractions (Represent, Compare)	Fractions represent a part (numerator) of a whole (denominator). Unit fractions have a numerator of 1 (1/2, 1/3). Fractions may represent a section (A); part of a group (B); or division (12 apples divided into 3 equal parts is 4) (C).

Complex ← Simple	**Equivalence of Fractions**	Recognition that some fractions that have different numerators and denominators represent the same part of a whole.	3/4 6/8 9/12
	Compose and Decompose Fractions	Compose: Recognition that fractions with like denominators can be combined into a single, larger fraction. Decompose: Recognition that fractions with like denominators may be represented by a sum of fractions in more than one way. 7/8 = 2/8 + 5/8 7/8 = 3/8 + 4/8 7/8 = 1/8 + 6/8	1/4 + 2/4 ——— 3/4
	Place Value (Read, Write, Represent, and Compare Whole Numbers and Decimals)	The location of a numeral determines its value, and that location determines if the numeral is representing a whole number or a part of a whole. For instance, the numeral 7 in the number 127.8 has greater value than the numeral 9 in the number 176.9 even though 176.9 is greater than 127.8 *and* the number 9 is greater than the number 7.	

Conceptual Understandings Related to Measurement and Data

Measurement and data skills developed in elementary years transition to more complex skills associated with domains such as statistics and probability in grades

6–8. Conceptual understandings learned through play in elementary years often represent the foundational basis for some very complex mathematical skills such as developing and utilizing a probability model for middle school learners. Phrases like "your block tower is taller than mine," "I have more cars than you," or "you're taller than me" integrate comparative concepts that are intricately related to number sense and measurement and data. In the primary grades, students' ability to understand and use comparative terms like *bigger than* or *more than* is the earliest phase of developing measurement and data skills. These very rudimentary concepts and skills must be thoroughly mastered for students to be able to access the more complex tasks, such as developing a probability model.

As students begin the journey toward mastery of early measurement and data learning outcomes, they must also develop the recognition that objects can be compared using various attributes. For instance, two cups can be compared by how tall they are (height), how wide they are (width), their weight (mass), or how much they hold (volume). Tables can be compared by height, length, width, area, and perimeter. These earliest comparisons are most often developed through exploring with uniform but nonstandard concrete units as opposed to the immediate introduction of standard units of measurement that are more abstract in nature. For instance, students can experience the comparative concept of measurement by using Unifix® Cubes or classroom plastic clips; the critical conceptual understanding is that the unit must be a uniform size. Student explorations with nonuniform units, under a watchful eye to ensure recognition of the importance of uniform size, represent rich learning opportunities. Students' conceptual understanding, fluency, and problem-solving experiences with measurement must be guided, from the use of nonstandard units to the use of the most efficient and appropriate standard units. For instance, it is not the most efficient choice to weigh a car using grams; kilograms should be used. By the time students progress beyond grade 5, their understanding of number and measurement must be fluid enough to use multiple strategies to find the volume of a rectangular prism or the area and circumference of a circle. While these examples may seem quite simplified compared to the curriculum expectations for grades six through eight, they may serve as appropriate entry points for students who require intensive, Tier 3 interventions to fill multiyear gaps in learning.

While comparing objects by various attributes is foundational for students to develop an understanding of measurement, an early understanding of data is also based upon comparison or sorting. The foundation for data analysis begins as students learn to sort objects based on various attributes. For example, they can sort buttons based on color and size. In each sorting scenario, students are beginning to identify data sets. Prompting questions like "Which group has more?" scaffold students connecting multiple elements of number sense and create an early data-analysis experience: Which group has more? Data-analysis skills that are developed in elementary grades often progress from sorting activities to data-collection experiences, where students identify the favorite class snack or favorite character in a book. These data-analysis tasks often begin with simplistic picture or bar graphs and answer questions such as, "How many more people like hockey than soccer?" and "How many people chose pizza or hot dogs?"

Like all mathematics learning, initial experiences with measurement and data will be most meaningful when investigations relate to the real world and involve concrete, hands-on use of tools. This may be even more impactful in an intervention setting where many of the students may have experienced years of frustration and disengagement with mathematics. Examples of connecting measurement activities may be as simple as approaching students with problems to solve such as "Which table will fit best in our room? What is our favorite healthy snack?" Measurement and data integrate students' understanding and applications of numbers into new, fluid contexts. Students' mastery of number sense is challenged and deepened within even the most basic context of measurement and data, requiring an application of knowledge of numbers and quantity to measure and analyze (for example, while fifty-seven is greater than forty-seven, it is not true that fifty-seven centimeters is longer or greater than forty-seven meters). Measurement integrates conceptual understanding of comparisons with conceptual understanding of number sense; each must be in place for students to demonstrate their understanding, application, and problem-solving skills. Table 4.3 (page 110) provides sample key vocabulary and concepts related to measurement and data that may be identified as critical for intervention for students in the middle grades.

Following are foundational measurement skills with which middle school students may have significant deficits that necessitate Tier 3 interventions.

Conceptual Understandings Related to Measurement and Geometry

Geometry, like all strands of mathematics, is composed of multiple, interrelated concepts that spiral in complexity. Lack of mastery of essential learning outcomes in the elementary school years often results in gaps in students' level of understanding that may interfere with their subsequent mastery of middle-grades-level curriculum. Instruction and learning related to geometry in the early grades are critical; the mastery of these concepts and terms serves as a prerequisite for future learning in multiple domains. Geometry in the elementary grades is focused on the study of 2-D

Table 4.3: Key Concepts—Measurement

	Concept	Description
		Conceptual Understanding → Procedural Understanding → Application
Complex ← Simple	**Measure Length Using Nonstandard Units**	As students begin to compare objects through measurement, concrete and familiar objects can serve as units if they are standard. For instance, the volume of a large bowl can be measured using a drinking glass as opposed to the abstract measurement of a liter; books can be measured using paper clips. Foundational understandings or big ideas related to measurement can be developed in this phase, including: <ul><li>The larger the unit, the fewer the number of units needed.</li><li>Objects can be measured or compared in a variety of ways. One bowl may have a larger circumference but a bowl with a smaller circumference may have a greater volume or weight.</li><li>Measuring with units allows us to more easily compare objects but this can only be done if the same units are used to measure both objects. A book that is fifteen paper clips long may not be the same length as a desk that is fifteen pencils long.</li></ul>
	Measure and Compare Lengths Using Standard and Nonstandard Units	As students progress with proficiency in measurement, they learn to use both standard and nonstandard units. While standard units are more abstract, they are necessary to efficiently communicate comparisons, especially about objects and distances that are not concrete. Standard units include all units that are defined and are universally accepted. A liter is a standard unit; a drinking glass is a nonstandard unit.
	Measure and Compare Length Using Standard Units	As students progress with proficiency in measurement, their work focuses on the use of standard units. The use of standard units is necessary to facilitate communication regarding comparisons, especially about objects and distances that are not concrete. As the use of standard measurement increases, students enjoy rich opportunities to employ the mathematical practices: model the mathematics, construct viable arguments, and choose appropriate tools. Standard units include all units that are defined and are universally accepted. Proficiency includes being able to convert units of measurement. Gallons and liters are standard units; a drinking glass is a nonstandard unit. A gallon can be converted to liters, and the conversion will hold true across settings; a gallon converted to a number of glasses may or may not hold true.
	Interpret, Model, and Compute Volumes of Rectangular Prisms	Recognize that the measurements of the length, width, and height of a rectangular prism determine the volume. Students should be able to model with manipulatives, create pictures, and use a formula.
	Understand Concepts of Volume and Relate Volume to Multiplication and to Addition	Through the use of concrete manipulatives and pictorial representations, students develop an understanding that volume is determined by repeated addition patterns and multiplication to determine volume.

and 3-D shapes. However, the study of shapes is more than the vocabulary or names associated with labeling or sorting shapes. Geometry in elementary grades includes the understanding of how shapes fit together, the spatial attributes of shapes, and their location and movement.

Geometry in the elementary grades establishes the precursors to fractional awareness; the partitioning of shapes leads to an understanding of unit fractions, and teachers at all grade levels can use this method to help students visualize fractions, including the need to find common denominators when adding and subtracting, and visual representations of all operations with fractions. The essential understandings or big ideas related to the understanding of shapes are complex and critical. These big ideas include the following.

- Properties of geometric shapes often are connected to measurements (two equal sides; right, obtuse, and acute angles; longer sides; and how far apart lines are at different points).

- Shapes are sorted by quantitative elements (number of sides, number of right angles) or qualitative elements (Will the shape tile; does it have curved lines or no curved lines?). (See figure 4.9 for definitions and visuals of tiling and nontiling shapes.)

- Lines and planes of symmetry are used to classify shapes.

- 3-D shapes often reflect the properties of the 2-D components (or individual parts) of the shape (see figure 4.10, page 112).

- Shapes can be used to represent parts of a whole and can be composed or decomposed to make larger or smaller shapes.

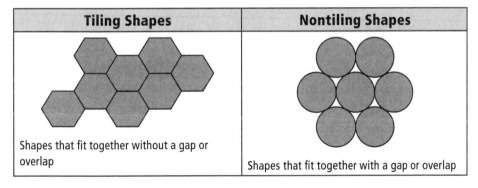

Tiling Shapes	Nontiling Shapes
Shapes that fit together without a gap or overlap	Shapes that fit together with a gap or overlap

Figure 4.9: Tiling and nontiling shapes.

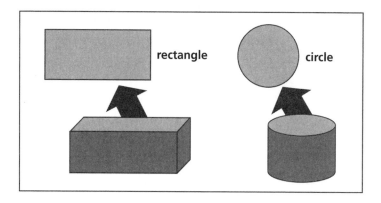

Figure 4.10: 3-D shapes reflecting 2-D components.

Proficiency within this domain includes significant attention to visual reasoning and spans all grades K–8 (Small, 2009a; Small, 2009b). This proficiency includes the ability to visualize individual components of both 2-D and 3-D shapes, to visually recognize the movement of shapes, and to be able to compose and decompose shapes. Table 4.4 provides sample key vocabulary and concepts related to geometry that may be identified as critical for intervention. Following are foundational geometry skills with which middle school students may have significant deficits that necessitate Tier 3 interventions.

Table 4.4: Key Concepts—Geometry

Concept	Description
Compose and Decompose Shapes	The ability to visually put shapes together and recognize the smaller parts of a more complex shape
Identify, Draw, Compare, and Classify Polygons	The ability to recognize, create, and sort closed 2-D shapes that have only straight sides **Polygon** **Not a Polygon**
Classify Polygons, Identify Symmetry	The ability to organize polygons by various attributes including number of sides, angles, locations, and lines of symmetry

Intervention Structures to Develop Number Sense

When planning for intervention, educators often ask us to recommend alternative programs or curricula. While many effective programs and curricula are available for instruction and intervention, research shows that teachers have more impact on student learning than programs. Effective teacher-to-student instruction including modeling, think-alouds, explicit direct instruction, and immediate corrective feedback more positively impacts the conceptual understanding and problem solving of students struggling with mathematics than programs do (NMAP, 2008; Slavin, 2014). Because teachers have such a significant impact on student achievement, our primary purpose in this work is to promote collaborative teacher capacity to meet student needs. To that end, we provide a sample intervention activity that targets a foundational skill with which middle school students may have a significant deficit: multidigit addition. We identified this skill as an example related to the development of number sense. As described earlier in the chapter, number sense includes the foundational, most critical skills necessary for all areas and domains of mathematics. Therefore, number sense should serve as one of the major elements of any mathematics intervention program. Please also note that effective use of these activities as part of Tier 2 or 3 interventions requires they be utilized within the context of effective instructional practices (modeling, think-aloud, explicit direct instruction, and immediate corrective feedback).

For our example, let's imagine that a universal screener of incoming sixth-grade students has identified a group of students that has scored well below grade-level expectations in mathematics. Further diagnosing and discovery reveal difficulties with multidigit addition, a skill that we would have expected and hoped that all students would master by the end of second grade. The school's leadership and RTI teams provide thirty minutes of intensive Tier 3 intervention four days a week to reteach multidigit operations to this group of students, starting with multidigit addition. These Tier 3 intervention supports are provided in addition to the differentiated Tier 1 instruction that they receive and to the Tier 2 intervention to which they have access, if common assessment evidence reveals the need.

Regarding multidigit operations including multidigit addition, teachers first assess students' ability to perform skills including estimating and justifying, calculating, imaging, and applying. The following figures provide examples of outcomes that indicate to the teacher whether an intervention is needed. The multidigit addition question used in this example to assess these skills is 26 + 18. After reviewing students' answers, teachers then provide needed interventions for these skills in a short lesson. Examples of how those interventions might be implemented appear after each figure. Finally, the student is asked to reflect on the effectiveness of the intervention. Teachers could utilize intervention strategies in the areas of measurement and data and geometry following a similar structure.

Estimate and Justify

Students should be able to use a grade-appropriate strategy to estimate and justify an answer (see figure 4.11).

Estimate (26 +18 =)	
Intervention Not Needed	30 + 20 = 50 (The student rounds 26 to 30 and 18 to 20.)
Intervention Needed	20 + 10 = 30 (The student is able to use the tens column only to estimate an answer.)

Figure 4.11: Teacher team's analyses to inform Tier 2 estimation needs.

The student's response will aid the teacher in helping the student move to a more sophisticated estimation strategy.

Using manipulatives or paper and pencil, ask:

- What does 26 look like?
- What does 20 look like?
- What does 30 look like?
- Does 26 look more like 20 or 30?

Calculate

Students should use a grade-appropriate strategy to solve the problem. They may show multiple ways or nontraditional methods to come to an accurate answer (see figure 4.12).

Calculate (26 +18 =)	
Intervention Not Needed	• Traditional stacked approach • Adding the tens first and then the ones • Adding 30 + 20 = 50 and then subtracting 6
Intervention Needed	• Errors in regrouping

Figure 4.12: Teacher team's analyses to inform Tier 2 calculation needs.

The student's response gives insight into his or her thinking and helps the teacher intervene with an appropriate lesson.

To assist this student with regrouping, we would simplify the question.

- Isolate 6 and 8 on the same piece of paper representing the ones place value.
- Once the student understands that 6 + 8 = 14, move the 1 to a second piece of paper (representing the tens place value), leaving the 4 behind.

Image

The student should be able to represent the mathematics using a sketch or diagram (see figure 4.13).

Image (26 +18 =)	
Intervention Not Needed	• Using a base-ten diagram • Using a number line to represent growth • Showing two smaller circles combining into a larger circle
Intervention Needed	• Student drew 28 dots and then 18 dots.

Figure 4.13: Teacher team's analyses to inform Tier 2 visual modeling needs.

To assist the student with imaging strategies, we would have the student group his or her dots into sets of ten.

Apply

The student should be able to explain how to use the mathematics in a real-life situation (see figure 4.14).

Apply (26 +18 =)	
Intervention Not Needed	"Two classes are going on a field trip, and we need to know how many seats we will need on a bus."
Intervention Needed	"I use this in math class."

Figure 4.14: Teacher team's analyses to inform Tier 2 application needs.

To assist the student with application strategies, we would have small groups of students discuss where they would see the number 26 in real life (hockey player's season total of goals, gallons of fuel for a trip, and so on).

Reflect

The student should be actively conscious of his or her learning (see figure 4.15).

Reflect (26 +18 =)	
Intervention Not Needed	Student is able to articulate what he or she found easy or hard, and what he or she could do next time to improve with justification.
Intervention Needed	Student is unable to justify response.

Figure 4.15: Teacher team's analyses to inform Tier 2 self-reflective needs.

To assist the student with application strategies, we would prompt the student to recall intervention effectiveness (Bird & Savage, 2014).

Technology Resources

While instruction has a greater impact on learning than programs, having access to the right digital resources can enhance our effectiveness. Computer-assisted instruction has a strong research base for improving students' computation skills—which is necessary and critical for accelerated learning. Teams should consider the following questions when considering an investment in technology-based resources.

- Do the hardware needs of the program match our current resources? Do we have the resources to upgrade our technology to meet the needs of the program recommendations?

- Is the program web based or server based? Do we have the bandwidth to run the program effectively?

- What are the initial costs and ongoing costs of the program? Is it a per-student license or concurrent license? Are the fees one time or ongoing?

- What training and costs would be required for effective implementation?

- Based upon review of our data, does the program specifically and thoroughly address the critical areas of need for our students?

- Do the language and vocabulary of the program match or nearly match the language and vocabulary of our classrooms?

- Does the program provide diagnostic assessments to target areas of need?

- Does the program monitor student progress and mastery?

- Does the program alert educators when students do not make sufficient progress?

- Does the program provide supplemental resources that the education team can utilize?

- Does the program provide a pilot or preview opportunity? If so, what feedback do students provide regarding program engagement and clarity?

We recognize that computer-assisted instruction can enhance our effectiveness in addressing student skill needs. However, it is important to keep in mind that technology is a supplement to our work as educators and should not be viewed as supplanting teacher instruction.

Conclusion

Teachers' understanding of the mathematics their students should master significantly impacts student achievement, and their understanding of conceptual and procedural topics and applications of mathematics empowers their ability to design tasks that engage students and help them connect to prior learning. Teachers' understanding of mathematics also allows them to develop questions that challenge and clarify student thinking and identify potential misconceptions. In addition, educators' thorough understanding of mathematics provides an instructional base to differentiate for students who are missing prerequisite skills and for those who are ready for enrichment.

Several factors impact our ability to frame mathematical learning so that students develop an understanding of the *why* and *how* behind mathematics to complement a proficiency with procedures. Each individual teacher's fluid understanding of mathematics is critical. However, we also have a collective responsibility to ensure our horizontal and vertical teams have clarity regarding concepts and related vocabulary.

Our understanding of the mathematics behind prioritized concepts directly relates to the instructional strategies that we employ to ensure that all students learn mathematics concepts, procedures, and applications at high levels—conceptually, procedurally, and with the ability to apply topics to real-world situations. The next chapter will provide guidance on the assessment and intervention practices that can ensure that all students in middle grades attain mastery of mathematics that keeps them on track to be future ready.

Chapter 5

Integrating Assessment and Intervention

Success in mathematics is a moral imperative and "algebra is a civil right" (Moses, 2001, p. 5). If educators do not believe in each student's ability to master mathematics concepts and procedures, then we would have to consider why we are bothering to intervene. We must accept that some students will simply require alternative strategies to learn, that not every student will learn the same way, and that some students will require additional time. We must also believe in our ability to teach every student. A teacher's sense of self-efficacy significantly predicts the achievement of students, and middle school teachers' beliefs in their abilities to teach mathematics lag far behind their beliefs in teaching reading well (Ashton & Webb, 1986; Bandura, 1993; Coladarci, 1992; Dembo & Gibson, 1985). We must also believe, and communicate to students, that mathematics achievement is not dependent on innate ability; work ethic, effort, perseverance, and motivation exert a significant impact on learning (Duckworth, Peterson, Matthews, & Kelly, 2007; Dweck, 2006; Seligman, 1991).

We must explicitly model and enthusiastically communicate the message that all students can master mathematics. We can and must influence a shift in thinking, from intervention as remediation to intervention as promoting additional opportunities for accelerating learning. The goal for all students, including students requiring intervention support, is critical thinking and problem solving. We must believe that all students can think critically and problem solve; deficits and gaps in skills do not mean that students cannot demonstrate deep levels of learning.

Assessment is a misunderstood concept and tool within our profession, and we must have the courage to challenge the ways in which assessment is used and the way it is viewed. Assessment *for* learning, also known as common formative assessment, is among the significant practices in which teacher teams can engage (Black

& Wiliam, 1998; Bloom, 1984; Meisels et al., 2003; Rodriguez, 2004). Assessment *as* learning, also known as students' self-assessment of their learning or self-reported grades, has been shown to be one of the most effective practices schools can commit to (Hattie, 2009).

There can be no PLC or RTI without timely evidence to inform timely responses (Buffum et al., 2009, 2010, 2012; DuFour et al., 2010). Schools should employ a combination of assessments, in addition to the common assessments described in chapter 2, to gather the evidence that will inform timely and targeted supports for students. These assessments include (1) universal screening, (2) diagnostics, and (3) progress monitoring.

For each type of measuring tool, we provide examples that staffs can use in grades 6 through 8. The use of assessments within the RTI framework is a critical component for informing instruction—the use of assessment for RTI is truly a demonstration of assessment *for* learning.

Universal Screening

Universal screening is a relatively efficient and impactful process in which schools can and must engage. There are efficient, easy-to-administer, and easy-to-analyze universal screening processes and measures that should be used with all students so that teachers know immediately if students are desperately at risk in mathematics. Universal screening in sixth through eighth grades should involve assessing knowledge of number, place value, and computational fluency involving whole numbers, fractions, and decimals. It should include application-based word problems in those same areas (Gersten, Jordan, & Flojo, 2005; Lembke & Foegen, 2009; Locuniak & Jordan, 2008; Mazzocco & Thompson, 2005; Methe, Hintze, & Floyd, 2008).

When screening reveals that students are at risk in these foundational areas, it is likely students will need intensive Tier 3 intervention to close the achievement gap. Number sense is fundamental and foundational to every concept in mathematics (Buffum et al., 2012; Weber, 2013). The areas of mathematics in which screening should occur include the following key concepts within numeracy and computation (Lembke & Stecker, 2007).

- Number writing
- Strategic counting
- Magnitude comparison
- Missing number
- Quantity array

- Number meaning
- Single and multidigit computation in all operations
- Word problems with all operations

Universal screening has become a buzzword and common practice of schools. Why do we administer screening measures? Do we act on the information that these measures reveal?

Screening measures assist teachers in identifying students who are significantly at risk for failure and difficulties before the school year begins. It is clear that students with significant deficits in mathematics (or literacy or behavior) will experience significant challenges as they continue with their schooling and transition to the adult world, even when teachers scaffold and differentiate access to content. Teachers scaffold and differentiate instruction so that students have access to content. Providing immediate, intensive supports in foundational prerequisite skills when screening reveals that such deficits exist is the critical additional step.

Screening requires that we gather information on all students regarding their abilities in literacy, mathematics, and behavior. Staff should never end their school year without a list of students who will require immediate, intensive supports in foundational prerequisite skills at the start of the next school year. We should lack universal screening knowledge only for students who are new to the grade level cohort in a school.

Educators can also administer universal screens at the start of the school year. These beginning-of-the-year measures provide timely information on where students stand in relation to benchmarks as well as a baseline against which to measure past and future performance. We should administer universal screening measures three times per school year (at the beginning, middle, and end of the school year), and we do this to identify students who will experience significant difficulties in the absence of immediate, intensive supports.

We *cannot* predict that a student will have difficulty learning about percents in a seventh-grade unit in February when he has otherwise understood concepts well. We *can* predict that a seventh-grade student without mastery of multiplication will have difficulty with percents. We *cannot* predict that a student will be confused about the appropriate signs of answers when adding, subtracting, multiplying, and dividing integers in middle school grades, even though that student understands other number sense concepts well. We *can* predict that a middle school student who does not possess absolute automaticity with basic, single-digit computation will constantly and consistently struggle when trying to learn mathematics. Screening aids us in identifying students significantly at risk so that we can provide the necessary, immediate, and intensive supports.

Examples of the types of guides that teams can utilize in their systematic efforts to immediately identify students in need of the most intensive mathematics supports are shown in figures 5.1–5.9. These tools could be used by teachers to screen (or even to diagnose needs of) students in any grade to determine if deficits exist in foundational skills.

> Ask the student to solve the following problems.
>
> 2 + 4 = _____   9 + 0 = _____
>
> 4 + 5 = _____   7 + 0 = _____
>
> 4 + 4 = _____   3 + 1 = _____

Figure 5.1: Single-digit addition.

> Ask the student to solve the following problems.
>
> 15 – 7 = _____   9 – 0 = _____
>
> 6 – 2 = _____   3 – 2 = _____
>
> 15 – 8 = _____   11 – 5 = _____

Figure 5.2: Single-digit subtraction from numbers up to 20.

> Ask the student to solve the following problems.
>
> 9 × 6 = _____   1 × 6 = _____
>
> 1 × 0 = _____   8 × 5 = _____
>
> 2 × 6 = _____   0 × 5 = _____

Figure 5.3: Single-digit multiplication.

It is important to remember that this type of screen is a broad tool and that students may be able to demonstrate proficiency without any real understanding of the concepts. RTI is not about merely checking off learning outcomes; it is about students having sufficient understanding to act as a structure or foundation for the next level of learning. Being able to complete a simple accuracy assessment may not indicate that students have sufficient understanding to scaffold to the next layer of knowledge. Rather than assessing surface skills many times, screening should ask students to use and represent a concept in a number of ways. These ways would include estimating, completing algorithms, providing pictorial representations, applying content to the real world, and reflecting on the process (Bird & Savage, 2014).

Ask the student to solve the following problems.

$48 \div 8 =$ _____ $0 \div 5 =$ _____

$45 \div 9 =$ _____ $28 \div 7 =$ _____

$18 \div 9 =$ _____ $20 \div 4 =$ _____

Figure 5.4: Single-digit divisors from dividends within 100.

Ask the student to solve the following problems.

241	682	541	190
+286	+309	+369	+739

213	367	563	666
+613	+332	+279	+210

Figure 5.5: Multidigit addition.

Ask the student to solve the following problems.

409	312	899	294
−302	−208	−353	−233

895	750	838	695
−782	−528	−272	−454

Figure 5.6: Multidigit subtraction.

Ask the student to solve the following problems.

582	800	299	179
× 71	× 31	× 43	× 24

120	806	876	138
× 37	× 54	× 26	× 76

Figure 5.7: Multidigit multiplication.

Ask the student to solve the following problems.

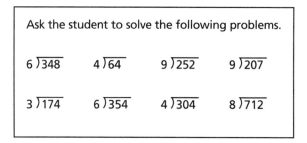

6) 348   4) 64   9) 252   9) 207

3) 174   6) 354   4) 304   8) 712

Figure 5.8: Multidigit dividends divided by single-digit divisors.

Ask the student to solve the following problems.

1. The grocery store sold 126 bags of Halloween candy this month. If each bag has 9 pieces of candy, how many pieces of candy did the grocery store sell in all?

2. The tree was measured to be 464 inches tall. The height of the tree needs to be reduced by 47 inches. How tall will the tree be after it is trimmed?

3. A warehouse holds 186 cases of pencils. Each case contains 5 pencils. What is the total amount of pencils the warehouse holds?

4. In the first week, Corinne's bean plant grew 25 centimeters. In the second week, it grew another 15 centimeters. How tall was the bean plant after the first two weeks?

Figure 5.9: Word problems with whole-number operations.

Diagnostics

Interventions, whether at Tier 2 or 3, will be most effective when they are targeted. Interventions are most targeted when they treat the causes, not the symptoms, of student difficulty (Buffum et al., 2009, 2010, 2012). This is not to suggest that diagnosing the exact causes of student difficulties is easy or quick. Determining the targeted supports that students need may be iterative—our first intervention efforts may inform and revise our initial diagnoses. Successfully targeting intervention so that students adequately respond to intervention is a process.

There are predictable reasons why students may experience difficulty mastering mathematics. Part of planning for instruction and intervention is anticipating errors that students may make. Due to this proactive planning, teacher teams can metacognitively model these errors, thereby demonstrating that learning mathematics is a process during which mistakes are normal and valuable. Teams can also collaboratively and proactively identify questioning that may uncover these misconceptions and the corrective actions to address them. Table 5.1 provides a sample of a common misconception that teams could identify, as well as a prompt and corrective feedback that teams could collaboratively create to uncover and address these misconceptions.

Table 5.1: Uncovering and Addressing Misconceptions

Common Misconception	Prompt to Uncover Errors	Corrective Feedback
$1/2 + 1/4 = 1/6$	How much money would you have if you had fifty cents and a quarter? How could we write this quantity as a fraction?	Use a rectangular model to represent 1/2. Use a rectangular model of the same length to represent 1/4. How could we add these two amounts?

Remember, the goal and necessity are for students to understand mathematics both conceptually and procedurally. In addition, students need to be able to apply their understanding. Determine in which of these areas (concepts, procedures, applications) the student is experiencing difficulties. When intervening, provide deliberate, explicit support to build student competencies and confidence. Using the example in table 5.1, build students' conceptual understanding through concrete objects (such as money) or rectangular models if students are making errors in the procedures. Provide scaffolds and steps if students can visually represent the concept but produce periodic inaccuracies when applying the procedures. Conceptual and procedural understandings are mutually reinforcing (Rittle-Johnson et al., 2001). Take advantage of this knowledge by ensuring that students possess both understandings for important topics. When teams establish that concepts and procedures are sound but that applying mathematics is proving to be a challenge, provide students with more explicit scaffolds and steps for problem solving, or modeling, with mathematics. A sample set of problem-solving steps is as follows.

1. Understand the problem.

2. Plan to solve with the problem in groups, individually, or both.

3. Agree on a solution to the problem.

4. Represent the answer graphically, visually, or in writing.

5. Defend solutions, critique the solutions of others, and make revisions to solutions.

Highly effective strategies exist for intervening in the area of mathematics (Baker, Gersten, & Lee, 2002; Gersten et al., 2008; Gersten et al., 2009; Hattie, 2009). The following are research-based recommendations for mathematics intervention.

- Deliver direct, explicit, and systematic instruction that models problem solving, metacognition, and step-by-step procedures.

- Provide students with immediate and specific corrective feedback in areas of need.

- Provide structured opportunities for students to talk with one another about mathematics.

- Provide scaffolded opportunities for students to verbalize their understanding.

- Visually represent and model mathematics problems.

- Focus intensively on properties of, and operations with, whole numbers.

- Focus on heuristics for solving word problems.

- Build automaticity with computation.

- Include motivational strategies within instruction, providing support that builds student confidence and competence with skills that are prerequisites of the target topic.

- Involve students in assessing their own work, goal setting, goal identification, and monitoring.

- Carefully select problems that address areas of need.

- Provide general problem-solving steps and multiple strategies.

- Use ongoing formative data and feedback, combined with targeted reteaching, to immediately address gaps in understanding.

- Provide feedback to students on the impact of their effort and growth.

- Use both computer-assisted and teacher-provided supports.

These strategies apply to core instruction and interventions. When common assessments reveal that students need more time and alternative approaches to master prioritized content, then Tier 2 supports must be provided. Intervention must address all three elements of mathematical learning: conceptual understanding, procedural understanding, *and* application. As stated earlier, the purpose of tiered instruction and intervention within the RTI framework is to ensure high levels of learning for all students. If interventions do not adequately address each of these elements in a scaffolded, measured way, we are not fulfilling the promise we make to learners within RTI. In fact, we are perpetuating the antiquated model of remediation or procedure-based intervention models. Our goal for intervention within an RTI context is to ensure student access to 21st century mathematics. Figure 5.10 illustrates the explicit sequence for intervention instruction. Intervention should explicitly and intentionally ensure that students master conceptual understanding, develop procedural understanding, and utilize those skills in an application-based (problem-solving) experience.

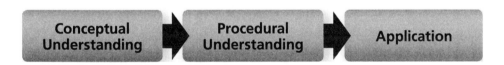

Figure 5.10: Scaffolded intervention design.

Diagnosis of Causes of Difficulties in Mathematics

The best interventions specifically target the causes and antecedents of student difficulties. If we can discover why a student is falling short of expectations, then we are more likely to help the student be successful. We too often provide general, one-size-fits-all programs that promise to improve entire content areas, such as *reading* or *sixth-grade mathematics*. While there may be a place for these types of interventions, students with significant deficits will respond most rapidly to supports that target the specific skills that are inhibiting their progress. The more diagnostic and specific the assessments, the more targeted, specific, and successful the intervention will be. Diagnoses of student needs can be time-consuming and are often conducted in one-on-one settings, but we are not diagnosing all students, only those in need of assistance. We diagnose the needs of students about whom we have concerns and questions to determine what specific supports are necessary.

We will learn so much about student needs through the process of intervening. A diagnose-prescribe-diagnose approach will provide at-risk students with immediate supports, can allow us to rule out possible explanations for poor performance, and can provide further insights into student thinking. There may be diagnostics that educators can purchase, but the most effective diagnostics come from information gathered through structured interviews with students.

One of the most powerful diagnostic activities is sitting with a student for twenty minutes to listen to his or her mathematical reasoning. To do this, the teacher presents a student with a set of problems from the appropriate grade level, one at a time, and observes. The teacher frequently asks why and what the student is doing, particularly, but not exclusively, when mistakes are made. Educators must trust themselves to accurately determine student needs, serve as diagnosticians, and use the information gained to inform supports for all students. Use the prompts identified in table 5.2 to diagnose student difficulties. To support educators in this powerful diagnostic activity, we have crafted these simple error investigations (EIs).

Table 5.2: General Error Investigation

Prompts	Notes
Present the student with a mathematics task.	
Ask the student to tell how to solve the problem.	
Ask the student to begin solving the problem.	
Ask *why* and *how* questions at every step of the problem-solving process.	
Ask the student to visually represent the problem (drawing, graph, bar model, array, rectangular model, or circular model).	
If necessary, present a visual representation *with* the student or *for* the student, and ask the student to interpret.	
Ask the student why the answer makes sense.	

Ask the student how the problem could represent something in real life.	
Ask the student to describe a similar problem.	
Examine where the student's problem solving seems to break down: • Sense of number • Computation • Use of the algorithm • Application of the procedures or steps • Understanding of the concept • Organization of the work (for example, answers are sloppy or imprecise)	

Source: Adapted from Hierck & Weber, 2014.

The knowledge gleaned from these sessions is invaluable, directly informing necessary supports for students. Learning targets in mathematics are much more specific to a grade level. Students who have deficits in number sense or computation are always going to be intensively at risk; however, we can collaboratively identify specific skills (such as regrouping when adding or subtracting) or processes (such as interpreting data) with which groups of students or individual students are struggling.

Error investigations (EIs) can help teachers and schools determine the types of supports that will allow students to positively and adequately respond to both initial instruction and supplemental interventions. We have the professional knowledge to meet students' needs. Our success is often limited by an inability to know the targeted areas in which students most require support.

Diagnosing student needs helps teachers to determine the types of supports that will allow students to positively and adequately respond to initial instruction and any supplemental interventions. Schools have the professional knowledge to meet students' needs. Success is often limited by an inability to know the targeted areas in which students most require support. Error investigations can be tailored to specific concepts. Table 5.3 (page 130) represents possible EI tasks for number, fraction, and decimal.

Table 5.3: Error Investigation for Number, Fraction, and Decimal

Possible Tasks	Notes
Ask the student to identify a whole number, a fraction, and a decimal.	
Ask the student to write a whole number, a fraction, and a decimal.	
Ask the student to represent a whole number, a fraction, and a decimal.	
Ask the student to create a story problem about a whole number, a fraction, and a decimal.	

Tier 2

Tier 2 supports involve more time and differentiated supports for students who have not yet mastered the priorities, as measured by regular common assessments. These supports are most commonly provided during *buffers*, time during which no new content is addressed. These buffers may be days that are proactively situated in the middle of units or between units, or they may be daily thirty-minute flex periods. During buffers, students are grouped homogeneously on needs that are revealed through common assessments. Students who have not yet mastered priorities receive support in a smaller group from teachers who have had the greatest levels of relative success, again as measured by the common assessments. Students who have demonstrated mastery of prioritized standards are presented with tasks that allow opportunities to learn at greater levels of depth and complexity. Other staff may join grade-level teachers to increase the teacher-student ratio during this flex time. To make optimal use of additional staff, schools may choose to stagger when each grade level has flex time.

The intent of Tier 2 is to ensure that students master prioritized grade-level content. We can predict that some students will require more time and alternative strategies and approaches to master the content initially addressed within Tier 1. Although we may not be able to identify which students will require this support or with which skills they will need alternative strategies and approaches, we can ensure that we gather the evidence required and that we secure the time, personnel, and resources to respond. Bloom (1968, 1984) recognizes this reality and demonstrates that 95 percent of students can achieve mastery of prioritized content when provided with

Tier 2 supports to supplement Tier 1 instruction. Tier 2 represents more time and alternative strategies to support students in mastering grade-level and content priorities. Strategies need not cost money, are not best found in a program, and can be identified through data analyses that reveal staff members who have had relatively greater levels of success helping students master specific skills.

At Myers Middle School, for example, the teacher team analyzed evidence from end-of-unit assessments that indicated that several students did not achieve an adequate level of mastery. Specifically, these students experienced difficulty translating the written description of the ratios into a numerical proportion. One of the teachers on the team had a greater level of relative success. She shared her strategy of interpreting word problems, which she called CUBS (circle, underline, box, state), in which students complete the following steps.

1. Circle numbers in both numeral and word form.

2. Underline operation words.

3. Box what the question is asking.

4. State the answer, leaving a space for the ultimate solution.

This teacher took the lead on providing intervention to students of various teachers on the team. She also used colored, concrete objects to represent quantities with the scenario as well as to write the proportion. When students receiving this Tier 2 intervention recompleted the corresponding items from the previously administered end-of-unit assessment, they demonstrated understanding at a significantly high level.

Tier 3

Intensive, targeted Tier 3 interventions are designed for students with significant deficits in foundational skills who have not responded to Tier 1 and 2 supports. These intensive supports should be provided in addition to Tier 1 and 2 supports. Given the constraints of the school's schedule and the immediacy and severity of student needs, Tier 3 supports may need to be provided, temporarily, in place of other important content, other than literacy and mathematics. Literacy and numeracy are foundational skills. We recommend that teams never supplant Tier 1 instruction in literacy or numeracy with Tier 3 interventions. Schools can creatively schedule these supports, providing them when students would otherwise be working independently. We believe that schools should not hesitate in providing Tier 3 interventions at the expense of other content; students with significant deficits in foundational mathematics skills are at grave risk of not graduating from high school, and the intensity of intervention should not be compromised.

Tier 3 supports should be adjusted to match student needs and revised until the student is adequately responding to intervention. We can and must identify students and their areas of deficit early. Once students with significant deficits are identified, we must provide intensive, targeted supports to begin closing gaps with a great sense of urgency; for example, we must end each school year having identified students requiring Tier 3 supports and arrived at an emerging understanding of the antecedents of their struggles, so that we begin the following year with intensive interventions the first week of school.

The goal of all students learning at high levels requires an all-hands-on-deck approach. *All* staff—special education, general education, teaching, administrative, and paraprofessional staff—must work together in creative and new ways to meet the needs of *all* students: students with Individual Education Programs (IEPs), students meeting current grade-level expectations, students exceeding current grade-level expectations, and students learning English as a second language.

The following are the foundational skills required for all students to understand mathematics, both conceptually and procedurally. The specific elements that comprise a number sense include but are not limited to:

- Subitize
- Magnitude
- Count
- Correspondence
- Cardinality
- Hierarchical inclusion
- Part-whole
- Compensation
- Unitize
- Computation
- Fractional awareness and proportional reasoning
- Word problems

While Tier 2 supports typically involve supporting students in the mastery of specific and prioritized grade-level skills, Tier 3 supports are typically necessary because of deficits in broader domains with foundational skills that were likely a prioritized standard years ago. Tier 3 supports typically require more time and more intensity.

In relation to Mr. Port's grade 7 class, the intervention team identified six students who needed intensive intervention to ensure their mastery of the prioritized foundational skill—using place value to compare three-digit numbers. Figure 5.11 provides a sample intervention lesson plan that illustrates increased explicit instruction, scaffolding of student ownership, integration of vocabulary, student and teacher think-alouds, and use of concrete manipulatives as they address the C-P-A instructional practice.

Mathematical Tier 3 Intervention Log or Lesson Plan

Intervention Focus: Using place value to compare three-digit numbers

Week of: October 15   **Interventionist:** Mr. Port   **Materials:** Base-ten blocks

Planning Guide

Monday	Tuesday	Wednesday	Thursday	Friday
Pose story problem that requires comparing or ordering three-digit numbers. Explicitly model ordering numbers with concrete, base-ten blocks. Use think alouds to include place-value review. Prompt group to tell teacher how to manipulate the base-ten blocks to represent number. Record new vocabulary in journal.	Review story problem. Prompt students as a group to tell you how to order numbers using base-ten blocks. Question prompts should require students to use place-value concepts. Change numbers in story; repeat. Students summarize process at end of class.	Review story problem. Students tell how to order numbers. Monitor student partner groups carefully. Correct misconceptions and use questioning to prompt.	Review story problem. Students self-talk through ordering numbers while you monitor and listen to individual students. Students write about process and example in journal. (Some students may use illustrations.)	Students complete common formative assessment for progress monitoring. Prompt individual students to think aloud one to three examples to ensure conceptual understanding.

Figure 5.11: Sample Tier 3 intervention plan.

Teams within the context of PLCs and RTI may consider utilizing a framework similar to figure 5.12 to guide their conversations. The questions and prompts within this framework can help collaborative teacher teams identify which students need extra time and support to master the prioritized standards. In addition, the framework helps those teams evaluate the effectiveness of instruction and intervention.

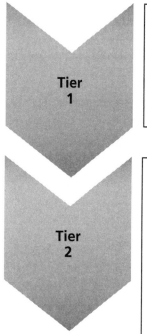

Tier 1

- Which students are not progressing well? Consider results of all assessment data including error investigation.
- Identify students whose needs can be met through differentiated instruction or reteaching. Monitor those students' progress to ensure Tier 1 support is sufficient.

Tier 2

- Group students with similar instructional needs. Consider across-grade or classroom grouping.
  - Conceptual misunderstandings
  - Procedural competency
  - Problem solving (or application)
- Identify or develop interventions that address targeted needs. Interventions should include C-P-A instruction, explicit instruction, and error correction. Interventions may also include computer-assisted instruction or strategies that are designed for English learners.
- Monitor progress (recommended every two weeks or as appropriate for skill or concept).

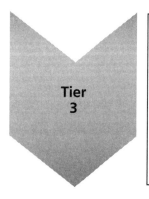

Tier 3

- Identify students who have significant gaps with foundational mathematics concepts or skills. *Significant* is relative to grade level, curricular needs, and students' response to Tier 2 intervention. School- or district-level teams discuss and define it.
- Provide intervention intense enough to accelerate learning. Provide daily instruction to supplement mathematical learning including conceptual understandings, procedural competency, and application of skills or concepts.

Figure 5.12: Tier framework for teams in an RTI and PLC context.

Progress Monitoring

There is no RTI unless we check on whether students are *responding* to intervention. We check on whether students are responding to supports through a process known as progress monitoring. The first type of progress monitoring in mathematics is designed for students who have been screened or determined to lack foundational prerequisite skills—students requiring Tier 3 supports. Monitoring progress to ensure that students are responding to the intensive interventions that are likely provided in these cases is most accurately and efficiently accomplished using the same type of assessments described in the section Universal Screening (page 120). The specific assessments that school teams select should, as much as possible, match the targeted skill with which the student is receiving support. These curriculum-based measurements (CBMs) must be at the instructional level at which the student is receiving support, which may be different from the student's current grade level. Figure 5.13 (page 136) provides a sample of progress-monitoring data for Mr. Port's Tier 3 intervention group.

For the second type of progress monitoring, prepackaged assessments are not widely available. When common assessments reveal that students require more time and an alternative approach to master prioritized standards, and when Tier 2 supports are provided, the most appropriate progress-monitoring assessment is simply another version of the common assessment that ensures students achieve mastery.

An obstacle that schools often face is an overall unwillingness to provide students with these opportunities. We are deluding ourselves if we express the belief that all students can learn without anticipating that some students will require additional time. We are not being honest with ourselves and with students and their families if we express that we are willing to do whatever it takes but not willing to provide students with (or do not *require* students to take advantage of) additional opportunities to prove their mastery. We are not implementing RTI if there is not a systematic expectation and opportunity for this type of progress monitoring.

Progress monitoring allows us to ensure that intervention is accelerating learning—that it is closing the achievement gap. Students at risk have no time to languish. If there is evidence that interventions are not resulting in adequate progress, we must further diagnose needs and the underlying causes of student difficulties and adjust supports accordingly. All students can learn—we must continue to problem solve and revise supports to ensure that high levels of learning are inevitable.

Student Data			
Student Name and Baseline Data (number correct / 20)	**Progress-Monitoring Data**	**Notes**	**Absences**
Lynn (3/20)	4/20	Can talk through with base-ten blocks with prompting	15th, 18th, 19th
Liz (7/20)	17/20	Moving to pictorial representation —little need for base-ten blocks	
Owen (4/20)	9/20	Needs hundreds chart	
Will (8/20)	12/20	Needs hundreds chart	
Dimitri (5/20)	14/20	Needs hundreds chart	
Lori (7/20)	20/20	Abstract, fluid understanding	

Figure 5.13: Student data.

We believe that a successful RTI will be measured by multiple indicators.

- The student is increasingly mastering grade-level priorities.
- The student is demonstrating more consistent mastery of specific prerequisite skills.
- The student's confidence and participation in mathematics are increasing.
- Work completion is improving.
- Performance on in-class quizzes and tests is improving.

Frequent analyses of progress-monitoring data are absolutely critical to ensure that students with deficits in mathematics are responding to intervention.

Conclusion

The more targeted the intervention, the better the intervention. When evidence indicates that students are not learning prioritized content at an appropriate level, we must first determine the causes of the difficulties. At Tier 2, our success at targeting interventions will be dependent on the quality of our common assessments and on our commitment to collaborative inquiry into student needs. When we screen to identify students with significant deficits in foundational mathematics skills, Tier 3 supports will be necessary. Determining the necessary focus of Tier 3 supports using the error investigations described in this chapter will be critical to ensuring that students respond in the timely manner that is critical to their future life chances. Students may also be identified as candidates for Tier 3 supports because they do not adequately respond to Tier 2 supports. No matter how we identify students, when they are found to have deficits in foundations, we must clearly diagnose their needs.

Once specific needs have been determined, intensive supports that match these needs must be provided with a sense of urgency. Tier 2 supports must involve more time and alternative strategies to learn the most highly prioritized standards. Tier 3 supports must focus on building the foundational skills with which the student has deficits.

To ensure that our supports are adequate and appropriately targeted to student needs, we must monitor student progress in a timely manner. We should expect with absolute certainty and confidence that students respond to interventions; high levels of learning are inevitable. We simply need to find the specific supports that students need. Success in mathematics is a critical prerequisite to students graduating ready for college or a skilled career.

Epilogue

The Promise and Possibility of Improved Mathematics Learning

Let's imagine and design an improved approach to application-based mathematics—to all students learning mathematics deeply and at high levels. What if there were no standards? What if there were no high-stakes testing programs? How would we design units of mathematics instruction? How would we teach? How would we use assessment to inform our future supports and to communicate to students where they are and what they need to do?

Expectations

We would start by nurturing a growth mindset among staff and students (Dweck, 2006). We would ensure positive and high expectations for students' abilities to learn at high levels. We would truly *believe* in our ability to ensure that all students learn at high levels. We would trust our colleagues because no single teacher can meet all the needs of a student. We would nurture staff capacities and collaboration. A sense of collective responsibility is critical for all students learning mathematics.

Focus

For decades, we have not had a viable mathematics curriculum (Gonzales et al., 2008; Marzano, 2003; Schmidt, McKnight, Cogan, Jakwerth, & Houang, 1999). When the amount of content that teachers are attempting to cover is not viable or doable, some students fall behind, some students become frustrated, some fail, and many lack opportunities to learn deeply, so that they can apply and retain knowledge. We must prioritize the most essential mathematics topics that students will need to apply in real-world situations. Students must understand these conceptually and procedurally so that they are ready for the next grade level or course, and ultimately, for college or a skilled career. We must also ensure that all staff members have a common interpretation of what it will look and sound like when students demonstrate mastery of the prioritized topics. What is the level of rigor, and what is the format in which students will demonstrate mastery? This amount of focus will lead to clearly articulated sequences of content progressions, both horizontally (within a school year) and vertically (from year to year). In designing a successful, robust, and balanced mathematics program, clarifying and concentrating instruction is foundational.

Evidence

Teachers and students need to know where they're going and where they are in order to determine what they need to do and how they're going to do it. Teachers and students need timely and specific evidence to inform the journey toward high levels of learning. The most effective and sound manner in which evidence can be created is through the use of common assessments (Fullan, 2005; Stiggins, 2007; Wiliam & Thompson, 2007). Tasks within common assessments represent the goals that all students will reach. They help teachers and students measure the concepts, standards, skills, and targets with which they are having success and the ones that require additional support. Students should self-report on their progress, guided through the use of scoring guides and anchor solutions to benchmark their performance. Teachers can collaboratively analyze student work to identify those targeted needs, the area in which the needs exist, and the staff members who may have had higher relative levels of success in helping students learn a given concept, standard, skill, or target. Assessments must join our knowledge of where we're going, where we are, and what we need to do to get there.

System

High levels of success cannot be based on a lottery-like approach—with some teachers employing these practices while others are not, and some grade levels making these commitments while other grade levels aren't. All teachers, grade levels, and schools must assume collective responsibility for all students learning at high levels, with high expectations for students, our colleagues, and ourselves. All teachers, grade levels, and schools must focus on a set of prioritized standards that is viable for all students to learn deeply. Teachers must serve as the model learner, guiding students to mastery with metacognition. Students must have opportunities to rehearse their learning and receive immediate and specific corrective feedback.

They must have opportunities to justify their own and their classmates' thinking and to critique the reasoning of others. Students must learn prioritized mathematics topics conceptually and procedurally and must have guided opportunities to apply mathematics and solve problems. We collectively concentrate to ensure that instruction is systematic.

There is virtually no single practice within education for which there is a more positive research base than common assessments. All teachers, grade levels, and schools must commonly assess with students as part of the learning process. Evidence from these assessments leads to urgent and regular Tier 2 supports so that students receive the additional time and alternative supports needed to learn the priorities.

We can predict that some students will need more support from time to time. We can also predict that some students will enter our classrooms with significant deficits in foundational mathematics skills. Schools can systematically screen or identify students who need intensive support in mathematics and provide intervention immediately by identifying the need at the beginning of the school year. Schools that ensure all students learn at high levels do not wait for students to fail for students to receive support. By raising expectations, focusing on depth over breadth, embracing evidence to inform learning, and proactively systemizing supports for students, we can ensure significant progress for all students' learning of mathematics. It will take courage. We know what we need to do, and we've recorded what we know in this book. We must now collectively commit to doing it.

References and Resources

Abbott, J. (2006). *Thesis 96: Curriculum for sustainability*. Accessed at www.21learn .org/archive/the-99-theses/thesis-96-curriculum-for-sustainability on January 11, 2014.

Ainsworth, L. (2003a). *Power standards: Identifying the standards that matter the most*. Englewood, CO: Lead + Learn Press.

Ainsworth, L. (2003b). *"Unwrapping" the standards: A simple process to make standards manageable*. Englewood, CO: Lead + Learn Press.

Ainsworth, L., & Christinson, J. (2006). *Five easy steps to a balanced math program for secondary grades*. Englewood, CO: Lead + Learn Press.

Allington, R. L. (2009). *What really matters in response to intervention: Research-based designs*. Boston: Allyn & Bacon.

Ashton, P. T., & Webb, R. B. (1986). *Making a difference: Teachers' sense of efficacy and student achievement*. New York: Longman.

Baker, S., Gersten, R., & Lee, D. (2002). A synthesis of empirical research on teaching mathematics to low-achieving students. *Elementary School Journal, 103*(1), 51–73.

Ball, D. L. (2005). *Mathematics in the 21st century: What mathematical knowledge is needed for teaching mathematics?* Accessed at www2.ed.gov/rschstat/research/progs /mathscience/ball.html on April 30, 2014.

Ball, D. L., Bass, H., & Hill, H. C. (2004). *Knowing and using mathematical knowledge in teaching: Learning what matters*. Paper presented at the 12th Annual Conference of the Southern African Association for Research in Mathematics, Science and Technology Education, Cape Town, South Africa.

Ball, D. L., Ferrini-Mundy, J., Kilpatrick, J., Milgram, R. J., Schmid, W., & Schaar, R. (2005). Reaching for common ground in K–12 mathematics education. *Notices of the AMS, 52*(9), 1055–1058.

Bandura, A. (1993). Perceived self-efficacy in cognitive development and functioning. *Educational Psychologist, 28*(2), 117–148.

Battista, M. T. (1999). The mathematical miseducation of America's youth: Ignoring research and scientific study in education. *Phi Delta Kappan, 80*(6), 424–433.

Bird, K., & Savage, K. (2014). *The ANIE: A math assessment tool that reveals learning and informs teaching.* Markham, Ontario, Canada: Pembroke.

Black, P., & Wiliam, D. (1998). Inside the black box: Raising standards through classroom assessment. *Phi Delta Kappan, 80*(2), 139–148.

Bloom, B. S. (1968). Learning for mastery. *Evaluation Comment, 1*(2), 1–12.

Bloom, B. S. (1984). The search for methods of group instruction as effective as one-to-one tutoring. *Educational Leadership, 41*(8), 4–17.

Brown, S. E. (2007). *Counting blocks or keyboards? A comparative analysis of concrete versus virtual manipulatives in elementary school mathematics concepts.* Accessed at www.scribd.com/doc/39698567/09 on April 30, 2014.

Bryant, D. P., Bryant, B. R., Gersten, R., Scammacca, N., & Chavez, M. M. (2008). Mathematics intervention for first- and second-grade students with mathematics difficulties: The effects of Tier 2 intervention delivered as booster lessons. *Remedial and Special Education, 29*(1), 20–32.

Buffum, A., Mattos, M., & Weber, C. (2009). *Pyramid response to intervention: RTI, professional learning communities, and how to respond when kids don't learn.* Bloomington, IN: Solution Tree Press.

Buffum, A., Mattos, M., & Weber, C. (2010). The why behind RTI. *Educational Leadership, 68*(2), 10–16.

Buffum, A., Mattos, M., & Weber, C. (2012). *Simplifying response to intervention: Four essential guiding principles.* Bloomington, IN: Solution Tree Press.

Bull, R., & Johnston, R. S. (1997). Children's arithmetical difficulties: Contributions from processing speed, item identification, and short-term memory. *Journal of Experimental Child Psychology, 65*(1), 1–24.

Charles, R., & Lobato, J. (1998). *Future basics: Developing numerical power* (Monograph). Reston, VA: National Council of Supervisors of Mathematics.

Coladarci, T. (1992). Teachers' sense of efficacy and commitment to teaching. *Journal of Experimental Education, 60*(4), 323–337.

Conference Board of the Mathematical Sciences. (2001). *The mathematical education of teachers.* Providence, RI: American Mathematical Society.

Darling-Hammond, L. (2010). *The flat world and education: How America's commitment to equity will determine our future.* New York: Teachers College Press.

Dehaene, S. (1997). *The number sense: How the mind creates mathematics.* New York: Oxford University Press.

Dembo, M. H., & Gibson, S. (1985). Teachers' sense of efficacy: An important factor in school improvement. *Elementary School Journal, 86*(2), 173–184.

Duckworth, A. L., Peterson, C., Matthews, M. D., & Kelly, D. R. (2007). Grit: Perseverance and passion for long-term goals. *Journal of Personality and Social Psychology, 92*(6), 1087–1101.

DuFour, R., DuFour, R., Eaker, R., & Many, T. (2010). *Learning by doing: A handbook for Professional Learning Communities at Work* (2nd ed.). Bloomington, IN: Solution Tree Press.

DuFour, R., & Marzano, R. J. (2011). *Leaders of learning: How district, school, and classroom leaders improve student achievement.* Bloomington, IN: Solution Tree Press.

Dweck, C. S. (2006). *Mindset: The new psychology of success.* New York: Random House.

Faulkner, V. N., & Cain, C. R. (2013). Improving the mathematical content knowledge of general and special educators: Evaluating a professional development module that focuses on number sense. *Teacher Education and Special Education, 36*(2), 115–131.

Fuchs, L. S., & Fuchs, D. (2007). A model for implementing responsiveness to intervention. *Teaching Exceptional Children, 39*(5), 14–20.

Fullan, M. (2005). *Leadership and sustainability: System thinkers in action.* Thousand Oaks, CA: Corwin Press.

Gersten, R., Beckmann, S., Clarke, B., Foegen, A., Marsh, L., Star, J. R., et al. (2009). *Assisting students struggling with mathematics: Response to intervention (RtI) for elementary and middle schools* (NCEE 2009–4060). Washington, DC: National Center for Education Evaluation and Regional Assistance, Institute of Education Sciences, U.S. Department of Education.

Gersten, R., & Chard, D. J. (1999). Number sense: Rethinking arithmetic instruction for students with mathematical disabilities. *Journal of Special Education, 33*(1), 18–28.

Gersten, R., Chard, D. J., Jayanthi, M., Baker, S. K., Morphy, P., & Flojo, J. R. (2008). *Mathematics instruction for students with learning disabilities or difficulty learning mathematics: A synthesis of the intervention research.* Portsmouth, NH: Center on Instruction.

Gersten, R., Clarke, B. S., Haymond, K., & Jordan, N. C. (2011). *Screening for mathematics difficulties in K–3 students* (2nd ed.). Portsmouth, NH: Center on Instruction.

Gersten, R., Jordan, N. C., & Flojo, J. R. (2005). Early identification and interventions for students with mathematics difficulties. *Journal of Learning Disabilities, 38*(4), 293–304.

Gersten, R., & Newman-Gonchar, R. (Eds.). (2011). *Understanding RTI in mathematics: Proven methods and applications.* Baltimore: Brookes.

Ginsburg, A., Leinwand, S., Anstrom, T., & Pollock, E. (2005). *What the United States can learn from Singapore's world-class mathematics system (and what Singapore can learn from the United States): An exploratory study.* Washington, DC: American Institutes for Research.

Gonzales, P., Williams, T., Jocelyn, L., Roey, S., Kastberg, D., & Brenwald, S. (2008). *Highlights from TIMSS 2007: Mathematics and science achievement of U.S. fourth- and eighth-grade students in an international context* (NCES 2009–001 revised). Washington, DC: National Center for Education Statistics, Institute of Education Sciences, U.S. Department of Education.

Gunderson, S. (2013, September 27). The postsecondary education investment [Web log post]. Accessed at http://blogs.reuters.com/great-debate/2013/09/27/the-postsecondary-education-investment on March 5, 2015.

Hanushek, E. A., Peterson, P. E., & Woessmann, L. (2011). Teaching math to the talented. *Education Next, 11*(1), 10–18.

Hattie, J. (2009). *Visible learning: A synthesis of over 800 meta-analyses relating to achievement.* New York: Routledge.

Hattie, J. (2012). *Visible learning for teachers: Maximizing impact on learning.* New York: Routledge.

Hiebert, J. (1999). Relationships between research and the NCTM standards. *Journal for Research in Mathematics Education, 30*(1), 3–19.

Hierck, T., & Weber, C. (2014). *RTI is a verb.* Thousand Oaks, CA: Corwin Press.

Higgins, L. (2008, May 27). Algebra I stumping high school freshmen. *Detroit Free Press.* Accessed at www.freep.com/article/20080527/NEWS05/805270337/Algebra-stumping-high-school-freshmen on September 14, 2014.

Hill, H. C., Rowan, B., & Ball, D. L. (2005). Effects of teachers' mathematical knowledge for teaching on student achievement. *American Educational Research Journal, 42*(2), 371–406.

Hunt, A. W., Nipper, K. L., & Nash, L. E. (2011). Virtual vs. concrete manipulatives in mathematics teacher education: Is one type more effective than the other? *Current Issues in Middle Level Education, 16*(2), 1–6.

Jayanthi, M., Gersten, R., & Baker, S. (2008). *Mathematics instruction for students with learning disabilities or difficulty learning mathematics: A guide for teachers.* Portsmouth, NH: Center on Instruction.

Jerald, C. D. (2009). *Defining a 21st century education.* Alexandria, VA: Center for Public Education.

Jordan, N. C., Kaplan, D., Olah, L. N., & Locuniak, M. N. (2006). Number sense growth in kindergarten: A longitudinal investigation of children at risk for mathematics difficulties. *Child Development, 77*(1), 153–175.

Jordan, N. C., Kaplan, D., Ramineni, C., & Locuniak, M. N. (2009). Early math matters: Kindergarten number competence and later mathematics outcomes. *Developmental Psychology, 45*(3), 850–867.

Jordan, N. C., Levine, S. C., & Huttenlocher, J. (1994). Development of calculation abilities in middle- and low-income children after formal instruction in school. *Journal of Applied Developmental Psychology, 15*(2), 223–240.

Kadosh, R. C., & Walsh, V. (2009). Numerical representation in the parietal lobes: Abstract or not abstract? *Behavioral and Brain Sciences, 32*(3–4), 313–328.

Kalchman, M., Moss, J., & Case, R. (2001). Psychological models for the development of mathematical understanding: Rational numbers and functions. In S. M. Carver & D. Klahr (Eds.), *Cognition and instruction: Twenty-five years of progress* (pp. 1–38). Mahwah, NJ: Erlbaum.

Kanold, T. D. (Ed.). (2012a). *Common core mathematics in a PLC at Work, grades K–2.* Bloomington, IN: Solution Tree Press.

Kanold, T. D. (Ed.). (2012b). *Common core mathematics in a PLC at Work, grades 3–5.* Bloomington, IN: Solution Tree Press.

Kanold, T. D. (Ed.). (2012c). *Common core mathematics in a PLC at Work, high school.* Bloomington, IN: Solution Tree Press.

Kanold, T. D. (Ed.). (2013). *Common core mathematics in a PLC at Work, grades 6–8.* Bloomington, IN: Solution Tree Press.

Kanold, T. D., Briars, D. J., & Fennell, F. (2012). *What principals need to know about teaching and learning mathematics.* Bloomington, IN: Solution Tree Press.

Kemp, K. A., Eaton, M. A., & Poole, S. (2009). *RTI and math: The classroom connection.* Port Chester, NY: Dude.

Kilpatrick, J., Swafford, J., & Findell, B. (Eds.). (2001). *Adding it up: Helping children learn mathematics.* Washington, DC: National Academies Press.

Kloosterman, P. (2010). Mathematics skills of 17-year-olds in the United States: 1978 to 2004. *Journal for Research in Mathematics Education, 41*(1), 20–51.

Kretlow, A. G., Cooke, N. L., & Wood, C. L. (2012). Using in-service and coaching to increase teachers' accurate use of research-based strategies. *Remedial and Special Education, 33*(6), 348–361.

Lembke, E. S., & Foegen, A. (2009). Identifying early numeracy indicators for kindergarten and first-grade students. *Learning Disabilities Research and Practice, 24*(1), 12–20.

Lembke, E. S., & Stecker, P. M. (2007). *Curriculum-based measurement in mathematics: An evidence-based formative assessment procedure.* Portsmouth, NH: Center on Instruction.

Light, J. G., & DeFries, J. C. (1995). Comorbidity of reading and mathematics disabilities: Genetic and environmental etiologies. *Journal of Learning Disabilities, 28*(2), 96–106.

Locuniak, M. N., & Jordan, N. C. (2008). Using kindergarten number sense to predict calculation fluency in second grade. *Journal of Learning Disabilities, 41*(5), 451–459.

Lyon, G. R. (n.d.). *Report on learning disabilities research.* Accessed at www.ldonline .org/article/6339 on June 17, 2010.

Lyon, G. R., Fletcher, J. M., Shaywitz, S. E., Shaywitz, B. A., Torgesen, J. K., Wood, F. B., et al. (2001). Rethinking learning disabilities. In C. E. Finn Jr., A. J. Rotherham, & C. R. Hokanson Jr. (Eds.), *Rethinking special education for a new century* (pp. 259–287). Washington, DC: Fordham Foundation.

Ma, L. (1999). *Knowing and teaching elementary mathematics: Teachers' understanding of fundamental mathematics in China and the United States.* Mahwah, NJ: Erlbaum.

Maccini, P., & Ruhl, K. L. (2000). Effects of a graduated instructional sequence on the algebraic subtraction of integers by secondary students with learning disabilities. *Education and Treatment of Children, 23*(4), 465–489.

Marzano, R. J. (2001). A new taxonomy of educational objectives. In A. L. Costa (Ed.), *Developing minds: A resource book for teaching thinking* (3rd ed., pp. 181–189). Alexandria, VA: Association for Supervision and Curriculum Development.

Marzano, R. J. (2003). *What works in schools: Translating research into action.* Alexandria, VA: Association for Supervision and Curriculum Development.

Mazzocco, M. M. M., & Thompson, R. E. (2005). Kindergarten predictors of math learning disability. *Learning Disabilities Research and Practice, 20*(3), 142–155.

McREL. (2010). *What we know about mathematics teaching and learning* (3rd ed.). Bloomington, IN: Solution Tree Press.

Meisels, S. J., Atkins-Burnett, S., Xue, Y., Bickel, D. D., Son, S., & Nicholson, J. (2003). Creating a system of accountability: The impact of instructional assessment on elementary children's achievement test scores. *Education Policy Analysis Archives, 11*(9).

Methe, S. A., Hintze, J. M., & Floyd, R. G. (2008). Validation and decision accuracy of early numeracy skill indicators. *School Psychology Review, 37*(3), 359–373.

Milgram, R. J. (2006). *The mathematics teachers need to know.* Presented at the Center on Instruction Mathematics Summit, Annapolis, MD.

Miller, S. P., & Hudson, P. J. (2007). Using evidence-based practices to build mathematics competence related to conceptual, procedural, and declarative knowledge. *Learning Disabilities Research and Practice, 22*(1), 47–57.

Morgan, P. L., Farkas, G., & Wu, Q. (2009). Five-year growth trajectories of kindergarten children with learning difficulties in mathematics. *Journal of Learning Disabilities, 42*(4), 306–321.

Moses, R. P., & Cobb, C. E., Jr. (2001). *Radical equations: Math literacy and civil rights.* Boston: Beacon Press.

Moyer, P. S., Bolyard, J. J., & Spikell, M. A. (2002). What are virtual manipulatives? *Teaching Children Mathematics, 8*(6), 372–377.

Mullis, I. V. S., Martin, M. O., Olson, J. F., Berger, D. R., Milne, D., & Stanco, G. M. (Eds.). (2008). *TIMSS 2007 encyclopedia: A guide to mathematics and science education around the world* (Vols. 1–2). Chestnut Hill, MA: Trends in International Mathematics and Science Study and Progress in International Reading Literacy Study International Study Center.

National Commission on Excellence in Education. (1983). *A nation at risk: The imperative for educational reform.* Accessed at www2.ed.gov/pubs/NatAtRisk /title.html on February 9, 2015.

National Council of Teachers of Mathematics. (2000). *Principles and standards for school mathematics: An overview.* Reston, VA: Author.

National Council of Teachers of Mathematics. (2006). *Curriculum focal points for prekindergarten through grade 8 mathematics: A quest for coherence.* Reston, VA: Author.

National Governors Association Center for Best Practices & Council of Chief State School Officers. (2010a). *Common Core State Standards.* Washington, DC: Authors.

National Governors Association Center for Best Practices & Council of Chief State School Officers. (2010b). *Common Core State Standards for English language arts and literacy in history/social studies, science, and technical subjects.* Washington, DC: Authors. Accessed at www.corestandards.org/assets/CCSSI_ELA%20Standards .pdf on July 6, 2012.

National Governors Association Center for Best Practices & Council of Chief State School Officers. (2010c). *Common Core State Standards for mathematics.* Washington, DC: Authors. Accessed at www.corestandards.org/assets/CCSSI _Math%20Standards.pdf on June 27, 2014.

National Governors Association Center for Best Practices & Council of Chief State School Officers. (2010d). *Standards for Mathematical Practice.* Washington, DC: Authors. Accessed at www.corestandards.org/Math/Practice on July 11, 2014.

National Institute of Child Health and Human Development. (2000). *Report of the National Reading Panel: Teaching children to read—An evidence-based assessment of the scientific research literature on reading and its implications for reading instruction—Reports of the subgroups* (NIH Publication No. 00-4754). Washington, DC: U.S. Government Printing Office.

National Mathematics Advisory Panel. (2008). *Foundations for success: The final report of the National Mathematics Advisory Panel.* Washington, DC: U.S. Department of Education.

Nelson, B. S., & Sassi, A. (2006). What to look for in your math classrooms. *Principal, 86*(2), 46–49.

Nelson, B. S., & Sassi, A. (2007). What math teachers need most. *Education Digest: Essential Readings Condensed for Quick Review, 72*(6), 54–56.

Newman-Gonchar, R., Clarke, B., & Gersten, R. (2009). *A summary of nine key studies: Multi-tier intervention and response to interventions for students struggling in mathematics.* Portsmouth, NH: Center on Instruction.

Northwest Evaluation Association. (2011). *College readiness linking study: A study of the alignment of the RIT scales of NWEA's MAP assessments with the college readiness benchmarks of EXPLORE, PLAN, and ACT.* Portland, OR: Author.

O'Brien, T. C. (1999). Parrot math. *Phi Delta Kappan, 80*(6), 434–438.

Okamoto, Y., & Case, R. (1996). Exploring the microstructure of children's central conceptual structures in the domain of number. *Monographs of the Society for Research in Child Development, 61*(1–2), 27–58.

Olkun, S. (2003). Comparing computer versus concrete manipulatives in learning 2-D geometry. *Journal of Computers in Mathematics and Science Teaching, 22*(1), 43–56.

Organisation for Economic Co-operation and Development. (2011). *Lessons from PISA for the United States: Strong performers and successful reformers in education.* Paris: Author. Accessed at http://dx.doi.org/10.1787/9789264096660-en on December 23, 2011.

Passolunghi, M. C., & Lanfranchi, S. (2012). Domain-specific and domain-general precursors of mathematical achievement: A longitudinal study from kindergarten to first grade. *British Journal of Educational Psychology, 82*(1), 42–63.

Patterson, K., Grenny, J., Maxfield, D., McMillan, R., & Switzler, A. (2008). *Influencer: The power to change anything.* New York: McGraw-Hill.

Perry, M., VanderStoep, S. W., & Yu, S. L. (1993). Asking questions in first-grade mathematics classes: Potential influences on mathematical thought. *Journal of Educational Psychology, 85*(1), 31–40.

r4 Educated Solutions. (2010a). *Making math accessible to English language learners (grades 3–5): Practical tips and suggestions.* Bloomington, IN: Solution Tree Press.

r4 Educated Solutions. (2010b). *Making math accessible to English language learners (grades 6–8): Practical tips and suggestions.* Bloomington, IN: Solution Tree Press.

r4 Educated Solutions. (2010c). *Making math accessible to English language learners (grades 9–12): Practical tips and suggestions.* Bloomington, IN: Solution Tree Press.

r4 Educated Solutions. (2010d). *Making math accessible to students with special needs (grades K–2): Practical tips and suggestions.* Bloomington, IN: Solution Tree Press.

Rapp, W. H. (2009). Avoiding math taboos: Effective math strategies for visual-spatial learners. *TEACHING Exceptional Children Plus, 6*(2). Accessed at http://journals.cec.sped.org/tecplus/vol6/iss2/art4 on April 30, 2014.

Reeves, D. B. (2004). *Accountability in action: A blueprint for learning organizations* (2nd ed.). Edgewood, CO: Advanced Learning Press.

Reimer, K., & Moyer, P. S. (2005). Third graders learn about fractions using virtual manipulatives: A classroom study. *Journal of Computers in Mathematics and Science Teaching, 24*(1), 5–25.

Reinhart, S. C. (2000). Never say anything a kid can say! *Mathematics Teaching in the Middle School, 5*(8), 478–483.

Riccomini, P. J., & Witzel, B. S. (2010). *Response to intervention in math.* Thousand Oaks, CA: Corwin Press.

Rittle-Johnson, B., Siegler, R. S., & Alibali, M. W. (2001). Developing conceptual understanding and procedural skill in mathematics: An iterative process. *Journal of Educational Psychology, 93*(2), 346–362.

Rodriguez, M. C. (2004). The role of classroom assessment in student performance on TIMSS. *Applied Measurement in Education, 17*(1), 1–24.

Rohrer, D., & Pashler, H. (2010). Recent research on human learning challenges conventional instructional strategies. *Educational Researcher, 39*(5), 406–412.

Rudmik, T. R. (2013). *Becoming imaginal: Seeing and creating the future of education.* Calgary, Alberta, Canada: Geenius.

Schmidt, W. H. (2004). *Papers and presentations: Mathematics and science initiative.* Accessed at www2.ed.gov/rschstat/research/progs/mathscience/schmidt.html on April 30, 2014.

Schmidt, W. H., McKnight, C. C., Cogan, L. S., Jakwerth, P. M., & Houang, R. T. (1999). *Facing the consequences: Using TIMSS for a closer look at U.S. mathematics and science education.* Boston: Kluwer Academic.

Schmoker, M. (2011). *Focus: Elevating the essentials to radically improve student learning.* Alexandria, VA: Association for Supervision and Curriculum Development.

Schoen, H. L., Fey, J. T., Hirsch, C. R., & Coxford, A. F. (1999). Issues and options in the math wars. *Phi Delta Kappan, 80*(6), 444–453.

Schoenfeld, A. H. (1987). What's all the fuss about metacognition? In A. H. Schoenfeld (Ed.), *Cognitive science and mathematics education* (pp. 189–215). Hillsdale, NJ: Erlbaum.

Seligman, M. E. P. (1991). *Learned optimism: How to change your mind and your life.* New York: Knopf.

Shaywitz, S. E., & Shaywitz, B. A. (2003). The science of reading and dyslexia. *Journal of American Association for Pediatric Ophthalmology and Strabismus, 7*(3), 158–166.

Siegler, R. S., & Robinson, M. (1982). The development of numerical understandings. In H. W. Reese & L. P. Lipsitt (Eds.), *Advances in child development and behavior* (Vol. 16, pp. 241–312). San Diego, CA: Academic Press.

Skinner, C. H., Turco, T. L., Beatty, K. L., & Rasavage, C. (1989). Cover, copy, and compare: A method for increasing multiplication performance. *School Psychology Review, 18*(3), 412–420.

Slavin, R. E. (2014). What works in teaching mathematics. *Better: Evidence-Based Education, 6*(1), 14–15. Accessed at www.betterevidence.org/issue-14/what-works -in-teaching-mathematics on November 25, 2014.

Slavin, R. E., & Lake, C. (2007). Effective programs in elementary mathematics: A best-evidence synthesis. *Best Evidence Encyclopedia.* Accessed at www .bestevidence.org.uk/assets/elem_math_Feb_9_2007.pdf on January 12, 2014.

Small, M. (2009a). *Big ideas from Dr. Small: Creating a comfort zone for teaching mathematics, grades K–3.* Scarborough, Ontario, Canada: Nelson Education.

Small, M. (2009b). *Big ideas from Dr. Small: Creating a comfort zone for teaching mathematics, grades 4–8.* Scarborough, Ontario, Canada: Nelson Education.

Stiggins, R. (2007). Assessment through the student's eyes. *Educational Leadership, 64*(8), 22–26.

Stigler, J. W., & Hiebert, J. (2004). Improving mathematics teaching. *Educational Leadership, 61*(5), 12–17.

Swanson, H. L., & Beebe-Frankenberger, M. (2004). The relationship between working memory and mathematical problem solving in children at risk and not at risk for serious math difficulties. *Journal of Educational Psychology, 96*(3), 471–491.

Texas Education Agency. (2012). *Mathematics Texas Essential Knowledge and Skills.* Accessed at www.tea.state.tx.us/index2.aspx?id=2147499971 on September 14, 2014.

Van de Walle, J. A. (1999). *Reform mathematics vs. the basics: Understanding the conflict and dealing with it.* Paper presented at the 77th Annual Meeting of the National Council of Teachers of Mathematics, San Francisco.

Van de Walle, J. A. (2004). *Elementary and middle school mathematics: Teaching developmentally* (5th ed). Boston: Allyn & Bacon.

Vygotsky, L. S. (1978). *Mind and society: The development of higher psychological processes*. Cambridge, MA: Harvard University Press.

Weber, C. (2013). *RTI in the early grades: Intervention strategies for mathematics, literacy, behavior and fine-motor challenges*. Bloomington, IN: Solution Tree Press.

Western and Northern Canadian Protocol. (2008). *The common curriculum framework for grades 10–12 mathematics*. Accessed at www.bced.gov.bc.ca/irp/pdfs /mathematics/WNCPmath1012/2008math1012wncp_ccf.pdf on December 22, 2013.

Wiggins, G., & McTighe, J. (2012). *The Understanding by Design guide to advanced concepts in creating and reviewing units*. Alexandria, VA: Association for Supervision and Curriculum Development.

Wiliam, D. (2007). *Five "key strategies" for effective formative assessment* [Research brief]. Reston, VA: National Council of Teachers of Mathematics.

Wiliam, D., & Thompson, M. (2007). Integrating assessment with instruction: What will it take to make it work? In C. A. Dwyer (Ed.), *The future of assessment: Shaping, teaching, and learning* (pp. 53–82). Mahwah, NJ: Erlbaum.

Winne, P. H., & Perry, N. E. (2000). Measuring self-regulated learning. In M. Boekaerts, P. R. Pintrich, & M. Zeider (Eds.), *Handbook of self-regulation* (pp. 532–566). San Diego, CA: Academic Press.

Witzel, B. S., Mercer, C. D., & Miller, M. D. (2003). Teaching algebra to students with learning difficulties: An investigation of an explicit instruction model. *Learning Disabilities: Research and Practice, 18*(2), 121–131.

Wurman, Z., & Wilson, W. S. (2012). The Common Core math standards. *Education Next, 12*(3), 44–50.

Yoong, W. K. (2002). *Helping your students to become metacognitive in mathematics: A decade later*. Accessed at http://math.nie.edu.sg/kywong/Metacognition %20Wong.pdf on April 30, 2014.

Zentall, S. S. (1990). Fact-retrieval automatization and math problem solving by learning disabled, attention-disordered, and normal adolescents. *Journal of Educational Psychology, 82*(4), 856–865.

Zimmerman, B. J. (2008). Investigating self-regulation and motivation: Historical background, methodological developments, and future prospects. *American Educational Research Journal, 45*(1), 166–183.

Index

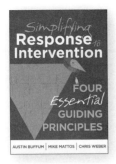

Simplifying Response to Intervention: Four Essential Guiding Principles
By Austin Buffum, Mike Mattos, and Chris Weber

The sequel to *Pyramid Response to Intervention* advocates that effective RTI begins by asking the right questions to create a fundamentally effective learning environment for every student. Understand why paperwork-heavy, compliance-oriented, test-score-driven approaches fail. Then learn how to create an RTI model that works.

BKF506

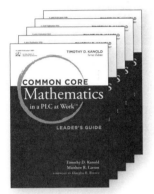

Common Core Mathematics in a PLC at Work™ Series
Edited by Timothy D. Kanold

These teacher guides illustrate how to sustain successful implementation of the Common Core State Standards for Mathematics. Discover what students should learn and how they should learn it at each grade level. Comprehensive and research-affirmed analysis tools and strategies will help you and your collaborative team develop and assess student demonstrations of deep conceptual understanding *and* procedural fluency.

Joint Publications With the National Council of Teachers of Mathematics

BKF559, BKF566, BKF568, BKF574, BKF561

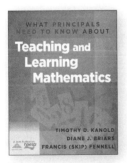

RTI in the Early Grades
Chris Weber

Explore why intervention and support for struggling students in the early grades are essential to student success. Teachers and support personnel will discover how to implement RTI-based supports in the early grades and learn what this prevention looks like. Find practical, research-based strategies to seal the gaps in student learning in grades K–3, identify students who need intervention, and more.

BKF572

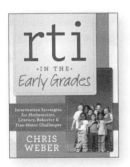

What Principals Need to Know About Teaching and Learning Mathematics
Timothy D. Kanold, Diane J. Briars, and Francis (Skip) Fennell

Ensure a challenging mathematics experience for every learner, every day. This must-have resource offers support and encouragement for improved mathematics achievement across every grade level of your school. With an emphasis on Principles and Standards for School Mathematics and Common Core State Standards, this book covers the importance of mathematics content, learning and instruction, and mathematics assessment.

BKF501

Solution Tree | Press

a division of
Solution Tree

Visit solution-tree.com or call 800.733.6786 to order.

WOW!

I liked how I was given
an effective, organized plan
to help EVERY child.

—Linda Rossiter, teacher,
Spring Creek Elementary School, Utah

PD Services

Our experts draw from decades of research and their own experiences to bring you
practical strategies for providing timely, targeted interventions. You can choose from a
range of customizable services, from a one-day overview to a multiyear process.

Book your RTI PD today!
888.763.9045

Solution Tree